Basic Concepts
for Managing
Telecommunications Networks
Copper to Sand to Glass to Air

NETWORK AND SYSTEMS MANAGEMENT

Series Editor: Manu Malek
Lucent Technologies, Bell Laboratories
Holmdel, NJ

BASIC CONCEPTS FOR MANAGING TELECOMMUNICATIONS
NETWORKS: Copper to Sand to Glass to Air

Lawrence Bernstein and C. M. Yuhas

COOPERATIVE MANAGEMENT OF ENTERPRISE NETWORKS

Pradeep Ray

NETWORK MANAGEMENT: Emerging Challenges and Trends

Edited by Shervin Erfani and Pradeep Ray

A Continuation Order Plan is available for this series. A continuation order will bring delivery of each new volume immediately upon publication. Volumes are billed only upon actual shipment. For further information please contact the publisher.

Basic Concepts
for Managing
Telecommunications Networks
Copper to Sand to Glass to Air

Lawrence Bernstein

and

C. M. Yuhas

Kluwer Academic/Plenum Publishers
New York, Boston, Dordrecht, London, Moscow

Library of Congress Cataloging-in-Publication Data

Bernstein, Lawrence, 1940-
   Basic concepts for managing telecommunications networks : copper
to sand to glass to air / Lawrence Bernstein and C.M. Yuhas.
     p. cm. -- (Kluwer Academic/Plenum Publishers network and
systems management)
   Includes bibliographical references.
   ISBN 0-306-46237-0
   1. Telecommunication systems--Management.   I. Yuhas, C. M.
II. Title. III. Series.
TK5102.5.B45 1999
621.382--dc21   99-42730
                                 CIP

2

ISBN: 0-306-46237-0

© 1999 Kluwer Academic/Plenum Publishers
233 Spring Street, New York, N.Y. 10013

http://www.wkap.com/

10 9 8 7 6 5 4 3 2 1

A C.I.P. record for this book is available from the Library of Congress.

Printed in the United States of America

7500   ASH-4893   10/11

For Natie. All the chapters are done.

Preface

This book is intended to orient people working in telecommunications within a hugely complex and constantly evolving industry. Each facet of telecommunications requires both formal and practical education to such a degree that the perspective of the whole can be lost in the intense specialization within one area. When that happens, individual decisions that might seem perfectly reasonable in a narrow context can sometimes be counterproductive to the larger enterprise.

Network managers need to understand the context and origins of the systems they are using. The evolution of the technology has led to compromises that are important to understand. Often, complicated solutions cannot be changed because of those compromises, and knowing the history of the systems can save the network manager from repeating historical mistakes. If the network manager can be shown why software, in its current state of development, is a fragile commodity, reasonable expectations and appropriate precautions can be employed to reduce the frustrations of network management. An awareness of the industry complexity can also help the network manager understand why simple management schemes that were once adequate for small networks will not scale up to larger networks. That networks are in a constant state of evolution and change is a fact of life can be welcomed with genuine exhilaration rather than fear and resistance.

Programmers need an appreciation of the application domain for which they write. They need an understanding of the reasons behind the interfaces they must satisfy and of the relationship of the software they build to the whole network. Programmers also need to understand that there is a subdiscipline of psychology called *human factors* that has amassed a body of experimental data on which they can draw. The data range from finely detailed parameters of human physiological responses to larger issues of human interaction with mechanized systems.

Sales representatives need to see the context into which their products must fit. They need to understand their customers' problems in order to integrate their products into an already intricately woven fabric of vendors and systems. Rarely will a sales representative encounter virgin territory. Therefore, it is important to understand what came before and how to meld new products with legacy systems.

The book is structured to allow a network or systems manager to either read through from beginning to end, as for a short course in telephony, or to choose only those areas where there is a need for study and practical application. The first four chapters describe the challenges of contemporary telecommunications, a view to the future, and a basic explanation of how networks evolved from twisted copper pairs to include silicon computer chips, fiber optics, and wireless transmission—(thus the book's subtitle: *Copper to Sand to Glass to Air*). Chapters 5 through 9 describe five critical aspects affecting system functioning which together form the "five pillars of success" for automation: network condition, system environment, software functions and features, database considerations, and human factors. The final chapter discusses economic impact and cost studies for proving in new systems. Appendix A gives the full compound terms for the abbreviations used in the text. Appendix B is a bibliography of other works that treat various aspects of the basic concepts introduced in this text in more detail and offer specific problem resolutions.

The authors gratefully acknowledge D. F. Snow for the use of his work in financial frameworks for evaluating major network investments, Bill Pugh and Gerald Boyer for their generous sharing of their research comparing broadband access alternatives, and Vinton Cerf for his charming explanation of the workings of the Internet. We very much appreciate the patient council of both Barry Boehm and Sandy Fraser. We also thank Manu Malek for his skillful editing and Harry Heffes and Bob Vetter for their helpful advice on early versions of the text. Shri Goyal of GTE Laboratories and Ken Lutz of Bellcore reviewed the draft version.

Lawrence Bernstein
Short Hills, NJ   *C. M. Yuhas*

Contents

List of Figures

List of Tables

Facing the Inevitable

Twisted copper pairs, silicon computer chips, fiber optics, and wireless transmission—*copper to sand to glass to air*—are the physical markers of the evolution of the telecommunications industry. Alexander Graham Bell's 1876 invention prospered for a century using a metallic telephone network technology with analog switches and copper transmission. Then in the 1970s, the development of the silicon chip changed all notions of capacities and capabilities. Computerization on a large scale arrived just in time to rescue an industry that was bursting with paper records. It brought the possibility of new services and automated networks. Today, fiber optic cable is gradually replacing copper to meet new demands for higher capacity services. Wireless services add the possibility of many combinations of copper, fiber, satellites, and towers.

1.1. UPGRADING THE ACCESS NETWORK

It is inevitable and desirable that incumbent local telecommunications carriers a upgrade the access network. There are 100 reasons for hesitating—it requires a large investment, the service demand is unclear, performance requirements are not fully standardized—but there are really only two choices. The carrier can drag its feet and lose business to competitors, or it can provide the faster service customers are demanding for Internet applications and telecommuting. As more people work from home or satellite offices, the demand for high-speed access networks increases. Companies will pay for their employees to have these services because of the inherent business savings, but they will pay only if those services are reliable, trustworthy, and competently managed.

There is no obvious best approach to rejuvenating the access plant. As happens with new installations, planners can be hurt by promises of future savings that evaporate when the system is installed. Yet it would be foolish to resist investing in high-speed access networks and new network management systems at just the time when they are key to the success of the business. What one can do is evaluate each unique offering and installation situation in light of principles that have supported past success.

1.1.1. Identify End User Solutions

Intuitive guesses by programmers or designers about how and why a system will be used are irrelevant. Disciplined surveys of user needs allow concrete objectives to be formulated. From these, criteria to judge the success of a new system or product can be generated. Objective measures of end users' ease of learning, error rates, use of features, and task completion can be used to determine further modifications.

1.1.2. Integrate Partial Network Management Solutions

Every new piece of equipment comes with some sort of system to manage itself. It is possible that the protocols for the new equipment differ from the protocols established in the rest of the network and therefore need to be integrated into what already exists. This is an annoying but well-bounded problem.

A more slippery integration problem surrounds the end users for whom careful system criteria were designed in the first step above. The network might have originally been designed with reliability having primacy. Introducing a new feature could change the way users employ the system and shift priorities, for example, from reliability to congestion management. Every integration of new equipment and features requires reevaluation in terms of the end users.

1.1.3. Estimate and Control

The cost of software development and system deployment must be well estimated initially and subject to formal control. There are tools and techniques to accomplish this, but most software development is still late, faulty, fragile, and expensive because the full arsenal of controls is not consistently applied.

1.1.4. Design to Accommodate

Network management systems must be designed to anticipate a plug and play environment, which is only partially realized today. There are problems inherent in migrating from circuit switching to voice-over Internet transmission, in shifting from ATM (Asynchronous Transfer Mode) to IP (Internet Protocol). Young industries are plagued with incompatibilities as different manufacturers and features jockey for position. Railroads warred over track gauge in the nineteenth century and the telecommunications industry is in the throes of protocol standards wars.

1.2. THE EVOLVING JOB OF NETWORK MANAGER

The industry is at the brink of combining three types of management—system, network, and services—to achieve totally integrated network management. Switching, transmission, and operations have given way to routing, transport, and intelligent network services as network design evolves. The story of the pressures that produced this change shows in part why systems look the way they do and also illuminates some of the reasons why people in the same business sometimes talk at cross-purposes.

1.2.1. Service Crisis Pressure

The telephone industry faced a service crisis in the 1970s. Telephone lines could not be installed in a reasonable time and reliability was an issue. The symptoms were first apparent in New York Telephone because that system faced the serious challenges of meeting the telecommunications needs of the world's financial centers. Soon it became clear that these provisioning complexities were a harbinger of problems that could become universal. Creating computer-based systems and centralized work groups to manage the network avoided a national crisis. Centralized Automatic Reporting on Trunks (CAROT) was among the first wave of operations systems; it replaced a manually configured trunk line test desk. Test results were reported to the testers who analyzed them and coordinated the work of technicians in the field to isolate and fix problems.

Meanwhile, time-sharing systems became popular and computer administrators began managing the lines and terminals that supported users as well as those supporting the computers themselves. The issues of queue management, system availability, and response time had primacy. To reduce traffic on these special-purpose networks, IBM introduced a terminal that could remember its last screen and compute differences, so that only those and not the entire screen had to be sent back to the computer. This also reduced traffic at the input/output computer ports, which were a chronic bottleneck. However, the computer administrators did not worry about managing the transport medium and the network people did not worry about the operation of the computers attached to their networks. One anecdote hints at how critical this narrow focus in both camps would become. A service provider installed a new high quality broadband line at a computer center with recovery mechanisms that were arbitrarily set just a little too long for an IBM computer. The line rarely failed but when it did, the computer stopped all of the sessions, and the network manager had to restart them all. A simple change to the retry monitor in the computer to make it compatible with the transmission line failure mode solved the problem, but only after each camp was made abundantly aware of the other.

1.2.2. Cost Pressures

During the 1980s, the need for cost reduction led to a second wave of operations systems powered by minicomputers that automated many of the craft functions in addition to providing the manual assistance typical of the first wave. A good example of this is the maintenance operations system that managed trouble tickets, did automatic testing and analysis, dispatched and then closed out problems without the need for intervention from people at the dispatch desk. The computer dealt directly and completely with the field technicians. The watchword for such systems became *flow-through*, a term coined at the time by D. L. (Lou) Casulli in the Operations Systems area of Bell Laboratories. Flow-through is the concept of executing a service order, which is a request for new service or a change to a system, with no human intervention beyond entering the initial request.

Now there were two critical elements preparing the field for new architecture: Mechanization was demonstrably workable and flow-through was desirable and cost effective. IBM introduced System Network Architecture (SNA), which allowed administrators to separate their physical networks from their logical ones and thereby let multiple applications share the same network. This was vitally important. The technology allowed computers and networks to be shared. The reduced cost of deployment made new applications in a shared environment much more attractive than stand-alone systems. It also allowed greater flow-through because networks were integrated.

1.2.3. Market Demands Force Network Management

Demand for new services and client/server computing continued to build in both the telephony and computer industries, but the limitation on the ability of telephone companies to deliver new services was the time it took to perform the "back office" functions. Competition from nonregulated providers, such as Teleport for example, was damaging particularly to the business customer sector. The nonregulated providers could deliver high-speed service in a matter of hours rather than the days it took a telephone company to set up initial service. The third wave of change stressed integrated and seamless operations as a goal, with a focus on both the network and computer aspects of network management. The ideal was illusive.

The trade press was filled with the struggles waged between telecommunications managers and MIS managers for control of networks that now carried both voice and data. Despite the more effective controls built into the operations systems and the systems used for managing the computer networks, they both suffered from being separate and apart from the actual network. Operations systems often gathered data from telemetry networks. The telemetry networks monitor, test, configure, and track every network parameter, but they are fragile and error-prone themselves.

1.2.4. Form to Function

The wave of the 1990s is to leverage the breakthrough technologies to allow managers to rise above the mechanics of networks to concentrate on the business purposes that need to be accomplished. Embedding management within the networks themselves is a shift from form to function. Thus, the proprietary time-sharing

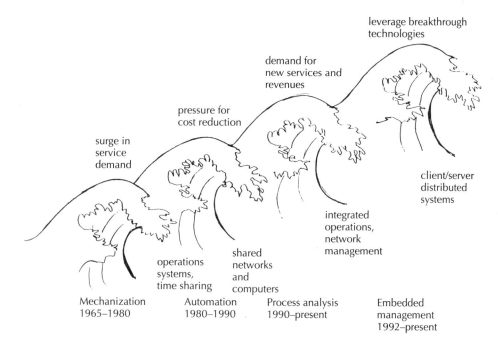

Figure 1.1. Four waves of operations systems management.

and transaction mainframe systems of the 1970s became the open minicomputer systems of the 1980s, which turned into the client/server distributed systems of the 1990s. Figure 1.1 summarizes the successive waves of methods for amplifying the power of computerization from simply mechanizing a purely manual task to entirely redesigning processes based on self-monitoring and self-diagnosing systems.

1.2.5. Terms of Confusion

Language has lagged behind these changes, with the result that computer people, telephone people, and service managers use the same vocabulary to mean different things and to pursue different goals. Indeed, even the term *network management* has different meanings. In telephony, it means the control of congestion in the network.[1]

Telecommunications is a discipline devoted to getting bits from point A to point B reliably and efficiently. These people measure costs and manage networks to optimize bit-delivery performance. They worry about security and billing. They assume the bytes will make sense of themselves if the bits arrive intact. They must juggle many networks: one carrying the message traffic itself, another carrying alarms and measurements from the network elements to the network management system and controls back to the network elements, one connecting the various network management computer systems together, and one bringing the information to the network manager's terminals and wall displays. Separate signaling networks provide for call management. In circuit switching, the route the call takes is established before the call begins. Internet and data networks add more layers and sometimes use special protocols for call management. Special routing protocols are used to establish routes.

For computer designers, network management means maintaining network availability and security.[2] Computer designers create local area networks and establish client/server hierarchies to transfer bytes from terminal to application. They worry about balances between server and network performance. They have little enthusiasm for discussions of point-to-point switching and performance because their assumption is that the bits will certainly get there. In these terms, the computer system is up if one terminal can send to one printer, even if another 999 terminals are down. By this philosophy, network management is "in-band," with systems detecting errors for other people to isolate and correct. For them, polling elements within the network or devices attached to the network to check status and detect problems are sufficient.

1.2.6. The Changing Nature of Work

One of the pioneers of the Internet, Marshall Rose, feels that the Simple Network Management Protocol (SNMP) is well focused and based on sound engineering principles, and that the failure to include desktop management to download application software and manage application configurations is a serious flaw in current network management. Not everyone shares this enthusiasm for SNMP. Telecommunication managers want out-of-band data and control paths, but SNMP relies heavily on in-band architectures. The use of client/server solutions has changed the nature of work by reducing the amount of rote work and democratizing the access to information. When employees become knowledge workers rather than paper shufflers, when entire new industries are created based on newly available access, the challenge is to manage these new systems.

Both the viewpoints of telecommunications and computer design are valid, yet neither is sufficient in itself. Vinton Cerf, the inventor of Transmission Control Protocol (TCP), could remark, "Most applications have no idea what they need in network resources or how they need to be managed."[3] The challenge at this point is to learn about the impact on resource reservation and management by modeling and experience.

1.2.7. The Changing Nature of Business Interests

Everybody wants to be in the telecommunications business. Worldwide, telecommunications carriers see competition from corporations that own extensive rights of way but are not necessarily in the communications business. In Japan and The Netherlands there are electric power companies, in France and The Netherlands, water companies, and in the United States and United Kingdom, gas line companies and railroads. Cable television companies everywhere view the lucrative telecommunications market as a possibility for corporate growth.

This competition fosters a need to offer innovative new services. From a business viewpoint, services must always be more, better, new, and improved. From a competitive viewpoint, liberalization of regulation and advances in technology have reduced the barriers to entry into the telecommunications business. A highly competitive environment results in one in which subscribers can select carriers based on the breadth and quality of services offered as well as the cost of those services. Broadband access architectures exist to meet the needs of emerging multimedia services and growth in the telephone plant. The capability of these broadband access architectures is further enhanced by an advanced services operations platform. To survive in this environment, carriers need a way to meet today's growth in the narrowband plant while they position themselves for the broadband future.

Figure 1.2 shows that customers can choose providers from many different levels. In a highly competitive environment, many types of providers vie to sell various categories of services to customers who may be large, medium, or small business networks or residential networks. For example, a customer may have several digital networks for transport to ensure reliability. Those who had such backup found it very useful during the April 13, 1998, AT&T frame relay problem.[4] Sometimes networks perform different functions; for example, companies will partition their voice networks from their e-mail messaging networks. In this case, they can buy individual networks from different suppliers. The disadvantage is that each

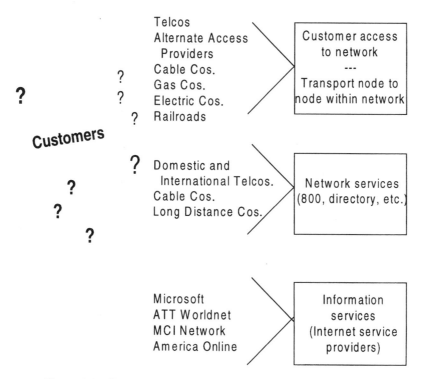

Figure 1.2. Customers choose providers from various categories.

comes with its own network management scheme. Other customers want their multiple function networks integrated under a full-service network concept with a fully integrated management approach. The controversy between these two extremes drives equipment vendors to build different solutions and makes it difficult to establish network standards. Other functions that customers desire are bandwidth on demand from network services and access to their small offices, home offices, and medium to large corporate offices. Access providers have tuned their offerings to meet particular needs, with the competitive differentiator being quality of service implemented through their network management scheme. In addition, customers demand mobility so that they can change their access to all services whenever they want to for telecommuting, traveling, and similar conveniences. To complicate the picture further, customers are seeking services tailored and individualized to their business needs.

1.3. FIVE PILLARS OF SUCCESSFUL AUTOMATION

Branch Rickey, General Manager of the old Brooklyn Dodgers, said, "Things worthwhile generally don't just happen. Luck is a fact, but should not be a factor. Good luck is what is left over after intelligence and effort have combined at their best. . . . Luck is the residue of design."[5] The five pillars supporting successful automation of telecommunications companies, as shown in Fig. 1.3, are the condi-

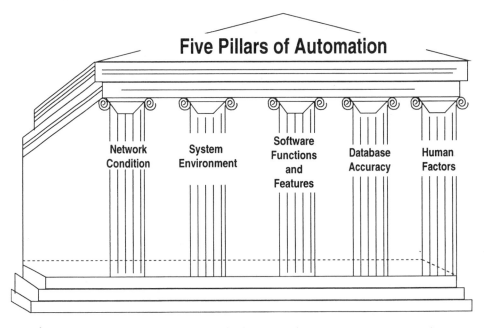

Figure 1.3. Five necessary elements of telecommunications company automation.

tion of the network, the preparation of the system environment, the software itself, the state of the data records, and the attention paid to human factors. Each must be fully understood and addressed. The tendency to assume that software is the main cost and primary focus is dangerous. Substantial investment goes into the other four areas also; for example, the cost of preparing the databases might be two to three times the cost of the software.

1.3.1.   Network Condition

The network plant must be in first rate condition and receptive to the monitoring and controls that a modern network management system must employ. Years of investment in information technology have made the telephone companies extremely productive users of new technology. From 1973 to 1983, productivity increased 6% annually. From 1965 to 1990, the number of lines doubled, traffic grew, and yet there was only a 14% increase in the number of employees. This is an unusually good statistic when compared across other U.S. industries. It resulted from the intense commitment the Bell System made to computer-based switching equipment, standardized business practices, and computer-based automation of manual functions. The breakup of the Bell System caused continuing investment to fall off in some cases, and the benefits arising from a commonality of processes were lost as local influences took precedence. The result is that telephone companies are not as successful today in squeezing costs, yet competition is forcing lower prices. The better-run carriers understand the effectiveness of investing in equipment and process analysis.

1.3.2.   System Environment

The second pillar, the preparation of the system environment, is least manageable. As each new network management system becomes part of the "system of systems" that describes how all of the parts of a carrier work together to provide service, the boundary conditions are the most difficult to characterize. Clear definitions and rules developed by standards organizations are typical efforts to manage the interfaces. Robust interfaces prevent the environment from becoming chaotic and supporting a standards policy is part of keeping the environment sound.

1.3.3.   Software

The software must have a rich set of functions, robustly designed so that the system can readily evolve as requirements change. Upgrading older systems is often so difficult that it is easier to start from scratch than to struggle with enhancing the original, but newer software should be better designed to allow easier change. The choice of a software supplier should take this into account, and initial cost cannot be the primary consideration.

1.3.4. Database Consistency and Accuracy

Keeping data correct and current is vital. *Consistency* means that a datum in one database appears exactly the same in all databases; *accuracy* means correspondence to the real world of specific equipment deployed, where it is, how it is used, and when it will be modified. Typically, carriers have multiple systems with some redundant data. Life would be easy for a Chief Information Officer if it were possible to have one common database feeding all of the operating disciplines. Data would be consistent because there would be only a single occurrence of the data. The data problem would resolve to a management issue of making sure the data corresponded to the real world. Unfortunately, telephone companies are too large, grow too fast, and are too diverse to make this a reality in any but the most unusual circumstances. Master plans sometimes have the goal of a common database, but when one element of the plan fails, local users pragmatically deploy a local system to solve the crisis. The quick fix becomes familiar and comfortable and is not replaced by the large corporatewide system. Thus, various systems coexist and the problem of data consistency among them arises. This all sounds like a simple matter of "bean counting," yet because of the enormity of the pile of beans, it costs a typical telephone company in the mid-1990s $2 per line per year[6,7] to keep data reasonably well.

1.3.5. Human Factors

The human factor is another key to successful automation of operations. People need to understand how their jobs will be changed and why the new method of operation is important to them and to the business. They must be trained and a new system should have been through sufficient human factors analysis and design to make it easy to use. Too often "should" does not happen and newness is rejected. People fear and resist change, especially when they have been doing a good job in the old way. The system designers need to guard against dehumanizing the work in their zeal to computerize. No system development can even begin, however, until current business practices are understood, documented, and standardized. Only with a baseline set of practices established can operations be analyzed and tasks be assigned to human or machine.

1.4. STRATEGIES FOR CHANGE

Flow-through requires the automation of many activities and judgments that were previously performed by people in order to reduce operations costs by minimizing the number of times a human has to enter data. Though the goal of achieving 100% flow-through is not possible because there will always be unusual circumstances or errors that require human judgment, minimum data entry should result in more consistent databases, though not necessarily more accurate databases in the sense that they reflect exactly the real world.

1.4.1. Engineering Flow-Through

Achieving high flow-through rates is a good first objective, but it is ultimately inadequate because it simply automates an existing manual process. The goal for the next generation is process reengineering. Getting good flow-through requires a thorough understanding of a manual process, then standardization and control of it. Reengineering that process means looking beyond to dynamic provisioning of resources and services, self-healing networks and proactive maintenance in anticipation of imminent problems.

The architectural considerations that are involved in delivering a completely integrated broadband access system tuned for maximum service flexibility and low operations cost underlie the choice of network equipment. Services that cannot be well managed cannot be offered. Fundamental plans should include intelligent network elements to interact with the Operations Support Systems (OSSs). OSSs are applications that manage various aspects of the network, such as rule-based systems to facilitate trouble reporting, analysis, and repair or process reengineering to allow real time service provisioning.

1.4.2. Providing Management Expertise

Capitalizing on the expertise of their employees should be a major competitive advantage for telephone companies in the process of reengineering and upgrading access networks. Companies with private networks have found it very hard to employ people with the right skills to staff network management centers. Countries in Asia and Africa with sudden demand for technology's gifts are discovering that hiring people with sophisticated skills in communications is a real difficulty. In these situations, automated network management systems that can operate with minimum staff may be the only viable option.

1.4.3. Providing the Connections

In North America and Europe, the copper plant used for voice telephone service is a liability. The copper network was consciously engineered during its first century to provide low fidelity voice transmission cheaply and efficiently. "Low fidelity" in this context means that the copper network would never provide the exacting fidelity of more sophisticated devices, but there was no need because the second part of the requirement was "voice" transmission. Pretty good fidelity over twisted copper pairs was good enough until high-speed data applications came along. As a result of various developments in transmission systems, there are now four types of connections or "loops" between end users and central offices:

- Nonloaded loops (without inductors to suppress noise) of twisted copper pairs up to 18,000 feet in length

- Loaded loops (with noise suppression) longer than 18,000 feet in length

- Derived loops with up to 12,000 feet of nonloaded twisted copper pairs connected to the remote terminal of a digital loop carrier system or a fiber to the curb system[8]

- Wireless avoids the individual connection to each user. A radio-transceiver box converts voice and data transmission to digital code that is sent at high radio frequency to a base station that connects to the switched network.

To move to a broadband plant, a telephone company must be prepared to absorb the value of its copper plant and the cost of replacing it. This is a formidable risk. The life of the copper plant could be lengthened by adding midband services. This strategy might satisfy current customers and keep their loyalty until broadband equipment can be installed. A significant investment in new plant administration systems would be necessary to do this because the physical data describing the equipment must be integrated with the logical information about how it is used, along with information about where it is located.

1.4.4. Justifying the Investment

Telephone companies must look to three areas to justify their investment in plant upgrades. The most important area is the current revenue base and the need to protect it from customer migration to more attractive services. The next is increased revenue opportunity from the ability to offer new and more reliable services. The third area is the improved productivity that would result from the reengineering required to upgrade. That the equipment would then be easier to maintain and provision would fund an additional portion of the investment. Complete replacement of deployed systems is unrealistic, but neither will customers tolerate continual change of their terminal equipment to accommodate stopgap measures. A realistic goal might be to modernize 10 to 20% of the access plant per year. The modernization would have to be done feeder by feeder, in the areas having the greatest demand for broadband services, rather than central office by central office as would be done in a copper network because that method would delay services to key customers. This could provide a workable compromise.

In any strategy, the only sustainable competitive advantage is customer service. People already rely on their telephones as a lifeline in innumerable ways that will only increase as possibilities for telecommuting, remote education, and remote health care become available. The systems to make this feasible must be compatible with existing networks and services, have open and efficient access to these networks and services, and must achieve a smooth evolution from early to more advanced systems.[9]

REFERENCES

1. Rey, R. F. *Engineering and operations in the Bell System*, 2d ed. (AT&T Bell Laboratories, Murray Hill, NJ, 1977).

2. Martin, J., with Chapman, K. K. *Local area networks: Architectures and implementations* (Prentice–Hall, Englewood Cliffs, NJ, 1989).

3. Bernstein, L., and Yuhas, C. M. "Network management and operations—The impact of technology," *The Froehlich/Kent encyclopedia of telecommunications*, Vol. 12 (Marcel Dekker, New York, 1996), pp. 477–503.

4. Bernstein, L., and Yuhas, C. M. "Chinks in the armor: Will a hybrid IP/ATM architecture protect the network from node failures?" *America's Network* **102:**13, 8–15 (July 1, 1998).

5. Rickey, B. "Faster than Jackie Robinson: Branch Rickey's sermons on the mound," *The New York Times* April 13, 1997.

6. Berkowitz, G. "Operations efficiencies of broadband access networks," *ISSLS '96 Proceedings* (Melbourne, Australia, Feb. 4–9), pp. 210–215.

7. Franklin, C. M., and Vogler, J. F. "Automated repair service bureau: Data base system," *The Bell System Technical Journal* **61:**6 (Part 2), 1131–1151 (July–Aug. 1982).

8. Hawley, G. "Systems considerations for the use of xDSL technology for data access," *IEEE Communications* **35:**3, 56–68 (March 1997).

9. Gillespie, A., Orth, B., Profumo, A., and Webster, S. "Evolving access networks: A European perspective," *IEEE Communications* **35:**3, 47–54 (March 1997).

Any Combination, Anywhere, Any Time

The popularity of the Internet and cultural trends such as telecommuting triggered a demand for multiple lines per customer in the United States, suggesting a soon to be worldwide growth in the need for telephone lines. Handling this sudden growth is no easy task. Telephone company planners are faced simultaneously with provisioning the explosion of lines linking homes and businesses with telephone company Central Offices (COs) and planning for a future where highly sophisticated interactive broadband services must be provided. John Mayo, a former president of Bell Laboratories, stated that the goal of network architecture is "to have access to voice, data and images, in any combination, anywhere, at any time—and with convenience and economy."[1]

What kind of network should there be? The choice seems to lie between a full-service network and a collection of networks dedicated to specific services like data or voice. The problem with building any new network is that the offered load or the applications are difficult to identify and cannot be characterized as readily as can voice traffic. Therefore, carriers are building a solution to an undefined problem. What does this imply about standardization or the lack of standards? Each new application changes traffic profiles, so the old probability models for traffic planning no longer work. A full-service network has to manage a mix of traffic, yet management tools are predicated on a model of voice communication. The basis of that model is that each communication is an independent event, an assumption now untrue in an environment where traffic is autocorrelated and bursty. This leads to phenomena like the pulsing congestion one finds on the Internet. Managing any portion of the network is complicated, but none more so than the access network known colloquially as "the last mile," which is the low-speed, low-capacity line that actually connects the home or business to the high-speed network. The speed at which new networks operate makes older kinds of analysis, which require time, too slow to use. Old solutions cannot be extrapolated to work in the high-speed world, which is both a worry and a challenge.

This chapter will discuss some questions that spring from a mandate to provide any combination of information, anywhere, any time, recognizing that the technology in this area changes so rapidly that by the time this is read, many more new questions will have arisen.

2.1. THE DOMINANT NARROW-BAND NETWORK

Narrow-band uses copper wires and loop electronics to carry voice communications. Telephone people call this Plain Old Telephone Service (POTS). A loop connects a telephone set to a particular switch in a CO containing one or more such switches. Before midband and broadband services were ever imagined, the telephone company loop plant was deliberately engineered to meet the following POTS specifications in order to provide the best service at the lowest cost with the most efficient sharing and reuse of each element:

- Optimize for low-fidelity voice communications limited to a bandwidth of 4 kHz

- Locate telephone set within 3 miles of its switch to limit loop load to 1300 ohms

- Send dial tone, the signal telling the customer that the switch is ready to accept the dialed number, to the telephone set within 3 seconds 95% of the time

- Optimize for average call length of 3 minutes during the busy hour

These engineering specifications allow all pairs to be treated interchangeably. An additional economy is the policy of *dedicated plant*, which asserts that it is most efficient to leave existing loops in place when a customer disconnects service because the next customer moving into the address will most likely need a loop with the same characteristics. When a customer requests a service that violates these principles, special engineering is needed to reconfigure the loop plant individually, case by case. This is called Special Service Provisioning and it is very expensive. Telephone companies incur a $125–$200 installation expense, excluding the cost of broadband equipment, to install a special service loop. By contrast, a POTS loop installation costs $25. A special service loop also costs a telephone company at least twice the installation cost yearly for administration because of its unique nature. As the telephone plant evolves, the wisdom of reengineered loops must be examined.

Telephone companies are caught in the dilemma of continuing to provision the last mile in the way they know best while recognizing that this way will not satisfy future needs. As Pugh and Boyer point out, forfeiting the economy of infrastructure sharing as the network nears the individual customer and investing in massive new Outside Plant (OP) construction instead of reusable electronic elements presents financial risk.[2] Telephone companies have experience with broadband technology because of the shared nature of high-speed trunks that connect switches to switches. This technology is practical and widely used to interconnect computers and to connect PBXs to the public telephone networks. However, the wholesale adoption of broadband technology into the loop plant changes the fundamental way the telephone company provides service and runs its business. Telephone engineers know that the customer demand for bandwidth to the home will grow, but they also know that it took 25 years to create the existing narrow-band infrastructure. Now that they have a primarily POTS infrastructure, telephone company planners struggle to justify investment in the broadband equipment, estimated at $500–$1500 per loop, in the absence of firm market "take" information. How to manage broadband loop networks economically also poses new problems.

There is some hope in a concept called *dynamic provisioning*.[3] It is based on a simple but powerful observation that call routes can be selected at the moment the call is made, rather than fixed during installation of service. If the access network itself chooses the route from a home to the CO for every call, even if it chooses exactly the same route each time, then complex, error-prone network management OSSs can be eliminated. When routes are fixed at installation, network management systems must track the already installed and future planned equipment and connections. The trunk plant, having built-in dynamic routing,[4] does not need to do this tracking. Bellcore's GR303 takes a step in dynamic provisioning. It automates connections between the OP and CO through time/slot interchanges which are available in new digital loop carriers. Broadband services can be feasible if network management is embedded because annual administration costs are reduced to levels close to those for POTS.[5]

2.2. MIDBAND SERVICES ARE VIABLE TRANSITIONS

Telephone companies are exploring the use of Asymmetric Digital Subscriber Line (ADSL) technology along with the older Integrated Services Digital Network (ISDN) technology to extend the life of the copper plant. Depending on the distance from the CO to a customer, ADSL can transmit 1.5 to 9 Mbps and return 16 to 640 kbps from the customer to the CO. ISDN can send 128 kbps in both directions.

2.2.1. Costs

ADSL can coexist with POTS on the same line and POTS can operate even when ADSL modems are turned off. The first costs of the ADSL equipment at the CO and at the customer's location can be as much as $1000 per loop in the current market. Projections anticipate bringing those costs to $400–$500 for mass deployment. The customer owns the ADSL equipment on the customer's site and the telephone company owns the ADSL equipment in the CO. For this reason, standards are critical in the manufacture of the hardware and software for this equipment. Standards are not yet available and therefore wide deployment has been delayed.

2.2.2. Engineering Considerations

Engineering of midband services is labor intensive and treats each loop as a unique entity for the following reasons:

- Each loop equipped with midband technology requires special reengineering to be certain that it can support the service.

- The loop must be examined to make sure that load coils, which are sometimes used to reduce noise levels, are not present.

- Range extenders, which are used in COs to overcome marginal signal conditions, must be removed.

- The length and gauge of the wire cable need to be checked to make sure that its inductance does not corrupt the digital signals.

- When digital loop electronics are in place, the plug-ins must be checked to make sure they are compatible with the ADSL or ISDN equipment.

These steps are similar to those needed for the expensive special service provisioning in narrow-band loops. They have the effect of undermining the uniform engineering principles that are fundamental to the economies of POTS by making each installation unique.

Rate-adaptive ADSL (RADSL) is designed to reduce these concerns. Some carriers plan to verify only the length of the loop and the absence of load coils. If these two criteria are met, they will use the RADSL to provide broadband service. RADSL functions like a dial-up modem; it senses the bandwidth capabilities of the loop and adjusts the transmission speed to fit the loop. RADSL will function for most loops, but at an unknown rate. The trouble is that the customer may gain little bandwidth benefit and there is no assurance that subsequent rearrangements of the loops will not degrade the bandwidth even further.

Another alternative is to condition all OP to handle splitterless ADSL, thereby pushing up the capability of POTS circuits. Splitterless ADSL operates at only a quarter of the ADSL bandwidth, but has the advantage that it appears to the CO as if it were a modem.

2.2.3. Administrative Complexities

Keeping track of where this expensive ISDN and ADSL equipment is installed and which loops are suitable for it requires new administration processes and systems. This administration is formidable for several reasons.[6] The administrative situation is similar to what existed before the engineered uniformity of POTS. Once ADSL or VDSL (Very-high-bit-rate Digital Subscriber Line) equipment is installed, the loop is unique. Dedicated plant concepts cannot be used in their present form. The engineering solutions mentioned in Section 2.2.2 assume standards that do not yet exist, so the customer site and the CO must use equipment from the same manufacturer in order to be compatible. A standard to make such equipment inter-changeable regardless of manufacturer will eventually be determined, but until that time, the carrier that wants the Small Office/Home Office (SOHO) market must do the administration.

An additional vexing practical problem is remote powering, which is needed to ensure lifeline service. When the power company has an outage, losing data communication is acceptable, but losing voice telephony isolates the customer and breaks one of the greatest values of telephone service. Power for Universal ADSL line cards at digital loop carrier huts or in COs and power on the line must somehow be supplied. ISDN solved these problems by using the CO powering plans.

Fiber is typically deployed to a digital loop carrier system from which customers are served over copper wires measuring 12,000 feet or less. The ADSL line card resides in the multiplexer in the digital loop carrier system. Power evolution is expected to keep pace with customer demand for Universal ADSL service. The demand for broadband service is estimated at 10% of all lines. This estimate is based on approximately 100 million lines in the United States, 40 million of which are PC owners. Of these, 30 to 50%, the SOHO group, will probably take ADSL. Between 10 and 15 million ADSL lines can be in use in a very few years.

If the critical design parameter of a 10% take rate should prove to be too low, the space and power budgets of COs will be inadequate. COs and mainframes built for POTS are a Procrustean bed for new technologies. To dissipate the heat generated by their higher power requirements, empty slots must surround ADSL lines and special air conditioning must be installed.

Unforeseen demand is a reasonable fear because experience with other products shows a sharp rise in demand over original estimates if service proves to be good. Once a powering scheme is established, new power management techniques must be added to the network manager's job to make sure service remains reliable.

If standards cannot be achieved and it becomes necessary for the carrier to own the equipment and rent it to the customer, the problem becomes even more complicated. This was done once before in the telephone industry. When local telephone companies used to rent telephone sets, they needed a system to keep track of them and other customer premises equipment in order to charge each month for each item. This tracking was done with paper records. Development of automated "left-in" administration computer systems stopped in 1984 when local companies no longer rented the customer premises equipment. With ADSL equipment in some homes but not in others, "left-in" administration systems would be needed again and in a far more sophisticated form than paper records. The presence of POTS splitters and multiplexes at the customer premises and at the CO must be managed to avoid extensive human testing to ensure compatibility for each installation.

2.3. BROADBAND MANAGEMENT

Broadband adds a new dimension of mediation and interpretation to the job of the network manager. The POTS functions of configuration management, fault location and repair, security, performance, and accounting are not enough. The existence of strong telephone industry standards in each of these areas and the ability to isolate points of trouble in the copper network made this job, if not easy, at least bounded and defined. Manufacturers devoted much money to the effort of making sure their offerings were compatible. The broadband network manager takes on the additional issues of interoperability and protocol mediation, quality of service, service agreements, and scalability.[7] Data networks present mediation problems because the technology outpaces the ability of the industry to define and enforce standards. For example, in the ADSL area there are new, nonstandard product offerings that have different protocols but give better performance, use less power, and cost less to administer. In a dynamic market, there is a constant tension between economy and standardization.

Narrow-band network managers are expected to track traffic, congestion, and error rates. The quality of the physical and logical networks is measured by bit errors and dropped frames. The performance of each network element can be assessed. The broadband network manager has those standards to maintain, but also tracks service performance against contracted goals, customer management of network assets, and performance of network elements as reported by their own built-in intelligence. The integration of the broadband network with existing support processes of electronic provision, customer care, application management, and accounting also fall to the broadband network manager.

2.4. BEYOND FLOW-THROUGH

One current goal of network managers is to achieve 100% flow-through because a by-product of that achievement is consistent, though not necessarily accurate, databases. The inability to share data in a common way adds cost by creating situations where large amounts of network information must be entered manually through expensive special translation software and coordinated among the various systems in the environment. Data often become inconsistent and unusable. In a typical trouble tracking database, as much as 25% of all records can be in error.

2.4.1. Reengineering the Process

But flow-through without process reengineering results in merely automating manual activities. The goal for the next generation of network managers is to reengineer processes based on the dynamic provisioning of resources and services with the help of self-healing networks and proactive maintenance in anticipation of problems.[8]

The objective is to create a completely integrated broadband access system, tuned for maximum service flexibility and low operating cost. These objectives are not mutually exclusive. The copper network achieved both objectives through standardization of equipment and a rate schedule that distributed costs not on the basis of use alone, but with a view toward universal service. An integrated broadband access system might also gain through standardization. Now, service differentiation and time to market are key to carriers providing broadband, and network management is one of the few areas where carriers can differentiate themselves.[9]

2.4.2. Combining Physical and Logical Data Models into a Single Model

Upgrading networks to broadband presents an ideal opportunity to combine today's separate physical and logical data into a single data model. Modern geographic data can be included and the costs of converting the physical paper records can be covered by avoiding the provisioning penalty incurred in using broadband within the constraints of today's loop administration systems. These data can be combined with the existing logical databases in a modern loop administration system to control costs. Without these changes, telephone companies can expect to pay high ongoing provisioning expenses for a network heavily populated with midband or broadband loops. Investments are needed now to make the inevitable growth of bandwidth in the loop plant affordable.

Unlike a copper access network built out of a varied mix of historical technologies, modern broadband access networks are characterized by fewer components with well-instrumented failure modes. The volume of traffic provides the money to pay for desirable protections, such as "lossless switching" where spare fiber is made available for insurance against failure on a cable. The intelligence built into these elements supports dynamic provisioning, automatic identification and activation,[10] performance monitoring, and fault isolation. The coordinated design of intelligent network elements and network management systems provides a unique opportunity to reengineer the way access networks and services are managed to gain efficiencies that would have been impossible in isolation.

2.5. OBSTACLES TO FIBER READINESS

Four obstacles have prevented embedding network management into the network and the result is an inefficient array of stand-alone systems.

2.5.1. Legacy Assumptions

The first is the "messy, real-life factor," or how to overcome the many assumptions that have led to the way networks are managed today. Entrenched organizations stand in the way of the introduction of technology that threatens to disenfranchise them. The tools and systems for network managers are hard to use and are not integrated, but, once laboriously learned, they are hard to abandon. Any OSS, regardless of its sophistication, is the intellectual descendant of the shoebox full of 3×5 cards that Alexander Graham Bell used to keep track of all of the details of the physical and logical network when he first strung wires in Boston. If this fundamental information could be embedded in the network elements, these shoebox databases could be tossed out. The network elements would automatically discover the data, store them, and update them. Just as the network elements determine the route for a long-distance call and the routing data are kept within the network, most other network management functions could be embedded in the network elements.

2.5.2. Funnel Factor

Next, there is the "funnel factor." Copper networks have naturally redundant pathways for information flows. No one line can carry much and there are frequent alternative choices to circumvent troubles. This is not the case with fiber networks. Trickles of data from far-flung reaches of the network become vast torrents of information when they are funneled into the common conduit of fiber. When an earthquake rearranges the geography where the fiber physically sits, a great many network nodes will suddenly drown in the backwash. Today's switches are too slow to handle the full speed and volume capacity of fiber, so the need for multiple switches acts as a series of "dams" to the rushing river of data. Even this small insurance disappears with the introduction of high-speed ATM switches. As the speed of the switches increases, they will handle more traffic, but without robust congestion management schemes, they too will become single points of catastrophic failure.

Current network management is inadequate if there is a flood of ATM data because of incompatible network management from different suppliers. SNMP cannot scale well to handle the volumes required, such as automatic topology features with huge network domains. Protocol analyzers will not be able to get to the troubled transport in time. Database "sniffers" will be incompatible with network analysis software. But this flood may not happen. The emerging technology of IP switching pours traffic directly into the network. IP traffic can be sent over Synchronous Optical Network (SONET) networks, which may even be replaced with dense wavelength-division multiplexing fiber optics. Just as ATM Quality of Service (QoS) issues were beginning to be defined and standardized, this change in technology occurred when it became clear that most traffic is not mixed, but mainly data. Because voice-over IP looks like data traffic, it became less pressing to spend money for ATM when there was no real traffic mix.

2.5.3. Software Reliability

The third factor is the increasingly complex software used to manage networks and the need for this software to perform without error. The software technologies of distributed computing, expert systems, software fault tolerance, and data management are a challenge both to the designers of network management systems and to their users. The disciplines for creating highly reliable, safety-critical systems will become important in the design of systems that manage services, systems, and networks on which the public depends.

2.5.4. Digital Factor

The "digital factor" is fourth. Digital systems do not fail gracefully; their internal health reports read, "OK, OK, OK, DEAD." Techniques to predict the failures of digital systems are in their infancy. The most successful ones measure the analog characteristics of a system and extrapolate the performance of the digital system.

2.6. THE INTERNET MODEL

What better model exists to handle large volume and withstand single points of failure than the Internet? An explanation by one of its inventors, Vinton Cerf, who provided the following analogy[11] to explain how the Internet works, will show both its strengths and weaknesses:

The Internet is a huge collection of over 100,000 computer networks that are interconnected around the world. Roughly 10,000,000 computers are "on" the Internet. These computers are sometimes called "hosts" because, historically, most applications were put up on large, central computers which "hosted" various services. Today, "hosts" can be clients OR servers or, sometimes, both. Networks are interconnected by special computers called "routers" whose job it is to "route" traffic from a source "host" to a destination "host" passing through some number of intervening networks.

The procedures used in the Internet to facilitate communication between computers are called, collectively, "protocols." These are conventions, formats and procedures that govern how information is organized and transported across the intervening networks between the source and destination computers.

Two of the key protocols used in the Internet are called "TCP" and "IP." Internet Protocol, or IP, is the most basic and on its foundation, the rest of the Internet sits. Transmission Control Protocol (TCP) is a "layer" above IP and provides services not available in IP. The two are usually referenced together as "TCP/IP".

"Traffic" on the Internet is made up of "packets" of data. Each packet has a finite, but not fixed length (content), a "from" address and a "to" address. It may be helpful to think of these packets as "electronic postcards" with all the features you already know about postcards. They are often called datagrams.

When you put a postcard into the mail slot, you have some expectation that the postcard will be delivered, eventually, to its destination. You don't know whether it will go by boat, by plane, by car or by train or perhaps all four. You are not really sure it will be delivered or, if it is, how long it will take. If you put in a number of postcards addressed to the same place, they may arrive in a different order than you sent them.

Internet Protocol packets are just like postcards, but about a hundred million times faster (assuming a postcard takes a day or two). When you send an Internet Packet into the Internet, there is no guarantee that it will be delivered. If you send several of them, they may be delivered out of order, they may even be accidentally replicated (something that doesn't usually happen to a postcard!). Internet Protocol provides what is called a "best efforts" communication service.

It is often a surprise for people to learn that guaranteed delivery is NOT a part of the basic Internet system. However, the next layer of protocol, containing TCP, makes up for the potential shortcomings of the Internet Protocol layer. The best way to understand what TCP does is to imagine what you would have to do if you were to try to send a novel to a friend, but the only way you could do it was to send it as a series of postcards.

First, you would cut up the pages so they could fit on a postcard. Then you would notice that not all the resulting postcards had numbers, and since postcards often arrive out of order, you think to number each postcard so your friend could put them back in order to make it easier to read the novel. Second, since you know that postcards may be lost, you would keep copies of each one, in case you have to send a duplicate to make up for a lost one.

How will you know if one is lost? Well, it would be convenient if your friend would send YOU a postcard every so often to say he had received all postcards up to postcard number N (for some value of N). On receipt of that postcard, you could discard the duplicates up to postcard number N that you had been holding. Of course, your friend's postcard might be lost, so you also need to have a kind of "time out" after which you start re-sending copies of post cards that have not yet been acknowledged. Your friend may not have sent any confirmation because he was missing some postcards.

If your friend receives duplicate postcards, he can easily deal with that since the cards are numbered, so duplicates can be ignored or discarded. Finally, you might realize that your friend's mailbox has a finite size. If you sent all the postcards of the novel at once, they might, by some miracle, all be delivered at the same time and might not fit in the mailbox. Then some would fall on the floor, be eaten by the dog, and you'd have to re-send them anyway. So it might be a good idea to agree not to send more than perhaps 100 at a time and await a postcard acknowledging successful receipt before sending others.

The Transmission Control Protocol does all these things to deliver data reliably over the basic Internet Protocol service. That's really all there is to it. Of course, I have left out some important other details such as routing (how do packets get to where they are supposed to go) and naming conventions.

Superimposing Internet protocols on ATM protocols can provide network robustness. *Connectionless communication* will drive the Internet to support private, reliable, voice communication. The Internet would need improvements, however, to handle voice communications well:

- Ability to interface to any public switched network and have compatible QoS measures

- Fast and reliable paths for data connections

- Remote concentrators to keep per customer costs reasonable; a special need will be testing copper loops to the customer

- Separation of routing and network operation from service provisioning (dynamic provisioning)

- Management of network congestion with an eye to recognizing symptoms that appear to be equipment problems but are in reality network traffic problems

Improvements are imminent to ease congestion with reliable and scalable directory servers.[12] Conveniences like voice mail, conferencing, call forwarding, dial-back, and whiteboards for collaborative efforts during teleconferences could make the Internet a formidable business tool. The combined forces of attractive flat-rate Internet service and the increasing complexity of network management to support number portability and unbundling make the trend to Internet voice attractive.

2.7. SELF-CONFIGURING AND SELF-MAINTAINING NETWORKS

The network itself must become capable of indicating what and where a problem is, calling attention to any special characteristics of the situation. Daugherty's intelligent access network controller provides a broadband asynchronous, time-division multiplexing capability for the OP.[13,14] Cross-connections between the switching office and any endpoint served by the switching offices are set up on a call-by-call basis. It can be accessed from terminals housed in neighborhood pedestals or in wire closets on customer premises. The controller establishes, in real time, the route that a call will take. It also contains adequate memory to store the needed administrative data, and it can be embedded in fiber networks. Housing similar self-identifying data in local area network hubs so that the data networks can be self-configuring can be practical. The controller maintains information about the end user, as well as any transmission peculiarities, so that each call can be tailored to the specific needs of the service, without sacrificing response, data, voice, or video quality. These controllers set up virtual circuits and paths for Internet messages using embedded network data.

2.8. SECURITY MANAGEMENT

Security management presents real problems. The issues of authentication, authorization, security, alarm reporting, and audit trails need attention. Because security breeches cannot be entirely prevented, network management systems had better be very good at detecting and tracking them.

Things will get even more complicated with SONET and Synchronous Digital Hierarchy (SDH) when the control and network management data are multiplexed on separate channels within the same physical circuits as the messages and signals. Understanding the use of these isolated channels must precede the use of ATM to multiplex all of these together on the same transport links.

2.9. CONGESTION CONTROL

Dr. Harry Heffes of Steven's Institute of Technology points out that traffic prediction techniques will be required to reserve bandwidth because the networks will not be able to react to surges quickly enough.[15] Traffic jams could become appalling on the broadband networks. Computer network designers, who are more interested in mapping the larger units of bytes, consider congestion management to be a brief passing cloud of mild concern, while telephony designers, who need each bit to arrive correctly, are consumed with avoiding it. Congestion is inadequately addressed in systems management. Messages describing component failures induced by congestion will quickly overload SNMP. Scalability is poor and network engineers need hands-on use of protocol analyzers to find and fix problems. This in itself is not bad, except for the time it takes and the custom-built arrangements required for protocol analysis to happen. This becomes a challenge for the telephony engineers to get the message streams nimbly to protocol analyzers.

Figure 2.1 suggests the complexity of the task of managing congestion in broadband networks. The problem is the amount of traffic that can flow in a short period of time and the approaches to the problem are two: preventive and reactive control. Preventive controls are designed into the network for the purpose of limiting the data coming into the network or making the network deliberately overengineered to handle more traffic than could ever be expected. For example, 40% is a typical planned utilization of TCP/IP over ATM networks and effective throughput rates can be as low as 34% as a result of interactions of the protocols.

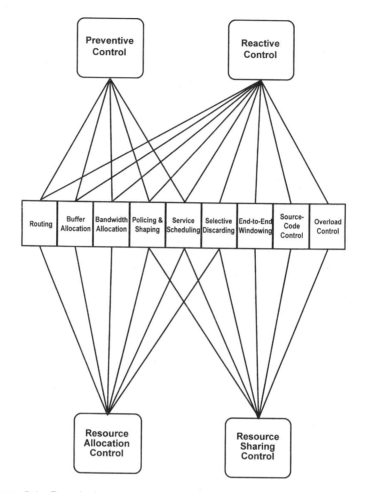

Figure 2.1. Complexity of congestion management in broadband networks.

2.9.1. Preventive Control

The first three areas of preventive control concern sizing and costs. Routing is the design of the layout of the links between routes and the choice of algorithm for the selection of links, which might be the shortest path, the least expensive path, or any other parameter that is critical to the service provider. Buffer allocation recognizes that the more buffers, the better the absorption of heavy traffic, within the limits of budget and size constraints. Bandwidth allocation is a prime area for deliberate overdesign of the network.

The fourth and fifth areas of preventive control concern methods of constraining the customer. Policing and shaping techniques place devices at the entrance to the network that will limit a customer to only the purchased service and will shape the traffic to the actual characteristics of the purchased service. Service scheduling assigns different levels of priority to classes of traffic.

2.9.2. Reactive Control

When the network senses that something is happening to threaten its efficiency, it takes steps to change what was done by the preventive control algorithms. Routing can be altered if there are focused points of overload, buffer allocation schemes can be modified, or particular customers might be shut off. If these methods are unsuccessful, there are additional techniques that may be employed.

Selective discarding allows specific packets to be dropped in specific services. For example, voice traffic could tolerate some dropped packets and still retain sufficient quality for intelligibility whereas data traffic between computers would be a very poor candidate for this control method. End-to-end windowing extends or shortens the window of information for breaking information into smaller pieces as more burstiness occurs in order to get more through the network without changing the routing algorithms. When a laptop user logs on to a network from a place distant from the Internet service provider, a special "foreign agent" contacts the service provider and relays security passwords and log-on information. If everything checks, the user has access to the Internet. The messages from the user and the responses might have to travel across the network and add traffic. Instead, a technique called *tunneling* can be used to encapsulate packets with source and destination information, therefore shortening the route the information must take. Source-code control is an extreme method that is not yet in common use, but will soon be available as a tool. It will be possible to download changed code to multiplexers and end devices to alter policing and shaping without taking the network offline, or dynamically change any algorithm. Overload control changes routes depending on the load offered to any route.

2.9.3. Resource Allocation and Sharing Controls

The resource allocation control assigns resources to services. The resource sharing control determines how services share capabilities. Figure 2.1 shows how they overlap for the already described areas.

2.10. CONFIGURATION MANAGEMENT

Configuration management is a major issue. Reliable, cost-effective, and easy-to-use backup and restoration systems are needed for broadband networks. When it comes to software, one needs to be able to "pack it and track it" as it goes among servers, clients, managers, and agents. The same holds true for data. *Data*, not *database*, management will be the issue for these complex networks. What or who will trigger work to begin once a new feature is installed or signal startup to begin after recovery from an outage when complex linkages exist?

Autodiscovery of clients on a TCP/IP network is powerful, but this feature does not scale well and has been used sparingly in telephony applications. A generalized autodiscovery feature is needed that embraces the concept that the network is the database. It is also needed at the services level. Hewlett-Packard's OpenView platform manages configurations across networked systems. Instead of making multiple changes every time a user is added to the network, the administrator can use one command to configure the user's password, e-mail account, and downloaded software. HP relies on a "synchronization" function to true up the actual state of the networked applications and computers with the administrator's databases. The problem of mixing the network's physical inventory with logical data in UNIX databases is difficult. For example, in a prototype that was built to extract information from a network element and write it to a relational database, the hardest part was getting the client protocol stack just right, especially in its interaction with the server's relational database. The database demanded versions of the protocol stack that could not be purchased for the client. Once the specific configuration worked, it worked well, but it was not robust to changes in the client or the server.

2.11. PLUG AND PLAY WITH TMN

Telecommunications Management Network (TMN) offers a method for integrating network management across different networks. Network providers and equipment manufacturers need to adopt software component approach, yet there continues to be a need for robust software technology to make this practical. The *New York Times* has stated, "there are dozens of tales of large development projects that have foundered badly, as the software industry has attempted to transform software development into an engineering discipline. . . . Object oriented software can be a big success, making it easier to scale products and promising much quicker development cycles. But getting there at the same time a company's culture is being transformed is a recipe for disaster."[16]

Unfortunately, there is neither evidence nor theory to suggest that the TMN middleware or a network management application built on top of it runs reliably, can be scaled up, or is robust to changes. The industry continues to rely on extensive testing and careful definition of the domain of use to ensure reasonable performance. Components based on object-oriented software offer the hope of having a 'plug and play' network management system, where the addition of another network element would only require plugging in its element management system to standard middleware. TMN promotes large-scale software reuse and modularity which would tend to lower system first costs and future maintenance costs. This is attractive, but stability cannot yet be ensured when standards lag behind new services.

The TMN concept of a manager and agents for the management functions reflects the physical separation of "managing" OSSs from the "managed" network elements. As this concept has become more formalized, it has come to represent the logical separation of functionality. Manager applications typically add value to network information in a specific domain through analysis and inference and usually need to have an integrated view of the network of managed objects. Agent applications, on the other hand, must have intimate knowledge of the elements of the network that they are representing. These network elements may be, for example, switches and multiplexors, or they may be other management systems that are acting in manager roles themselves for other domains or other views of the network objects. TMN provides a framework for creating managers and agents and simplifies the software development that can be driven by the object definition phase.

Figure 2.2 locates the logical function layers of TMN middleware between the application and the equipment. TMN is the connective software between applications and the wireless, broadband, POTS, and private networks. The TMN middleware is divided into two layers, as shown by the dashed line in Fig. 2.2.

The higher layer, customer database, services manager, and network database, consists of network management functions that the OSS needs to control the network itself. Services management is a technique introduced by TMN. For example, if a customer wants call waiting, the customer can, by making the request, update the database and instruct the network to update the switch database to reflect the new service. A critical part of the network is the network database, which contains descriptions of the configuration and pending changes, including additions and changes to equipment. Before TMN, this information was held in the

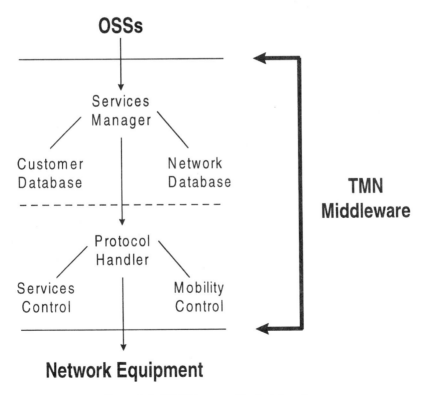

Figure 2.2. TMN layers of logical functions.

individual OSSs. The customer database is integrated into the services manager, thereby providing a coupling through the services manager with the network database. In the past, some customer information was mixed with network data, which created grave difficulties as far as accuracy was concerned. The example of call waiting was a simple one. The customer database is complicated by other factors such as tariffs, and the network database is complicated by the need to relate the equipment in one area to that in COs in other areas. The staging problem is one that TMN handles well in phasing in new services.

The lower level, services control, protocol handler, and mobility control, is the intelligent network logical layer. This provides the functions of security, personalization, and mobility. Personalization means how the network looks to particular business customers as they configure the network to meet unique business needs. Mobility means the customers are able to change where they appear in the network, for example, with number portability when their carriers or base locations change. The important new feature of this layer is the protocol handler, which translates information from the services manager in the layer above. The partitioning of functions permits different databases to use the same security approaches and mobility approaches within defined TMN standards.

There is a concept within TMN of a Management Information Base (MIB) that contains object class definitions, object instances, and attribute data for the objects contained in the data model that defines the network. This information needs to be stored and managed to be readily accessible to manager and agent functions. Providing these data store and management facilities can best be handled by using a TMN component-based approach. Because the TMN standards are written using object-oriented specifications and because the object-oriented technology promises more cost-effective and faster solutions, the next-generation OSSs are modeling the network as a collection of objects.

2.12. RECAPPING THE SITUATION

The copper network has been honed by 25 years of continual improvement to do low-fidelity voice communication well and reliably, but customers want much, much more than that. Dynamic provisioning is the right idea for reducing operations cost. Carriers could add some "sand" to the mix in the form of various digital loop electronics to make a transition through midband services, though this plan would require the greatest investment in network management. Or, carriers could go directly to broadband, either by totally reengineering the access network with the "glass" of fiber or, if the obstacles are too difficult to overcome, by leaping onto the Internet.

2.12.1. Considerations

Both broadband alternatives present the same set of considerations:

- Configuration and maintenance must be internal functions (like Daugherty's intelligent access network controller or others).

- Need to share transport and routing equipment requires standards.

- Security, congestion, and protocols must be mediated.

- Configuration management focus must be data packing and tracking.

- Logical separation of functions is necessary to handle complexity (TMN).

2.12.2. Envisioning an Ideal Network Management System

Figure 2.3 suggests what fully realized broadband management might be. From one point, any service, any system, any network element could be reached to give automated, near-real-time service activation, and continuous trouble surveillance. The enabling technology for this achievement would be an integrated data model built on descriptions embedded in objects, synchronized with network elements. The result would be significant savings over present methods.

An ideal network management system would detect anomalies from any source. Rather than dump a glut of raw bits, this system would digest data and transform them into information, presenting the network manager with only what was needed to resolve the problem. A manager could view a display well designed for easy human use, showing all elements of service operation from server queues to transport error seconds. The manager could point to any service, system, or network element and expect graphics of decoded messages being passed or filling a buffer. The manager could take performance measurements, then instruct a service

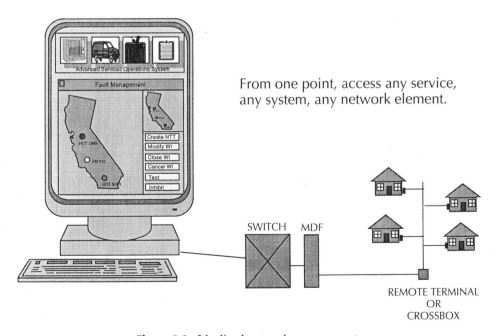

Figure 2.3. Idealized network management.

organization to make changes in its operation. Finally, from the same display, the manager could inject a test message into the application and trace it through the distributed system, either by allowing it to choose its path or by forcing one. This integration of boxes, links, elements, queues, utility software, and data managers is within current technological capabilities.

REFERENCES

1. Heldman, R. K. *Information telecommunications: Networks, products and services* (McGraw–Hill, New York, 1994), pp. 9–10.
2. Pugh, W., and Boyer, G. "Broadband access: Comparing alternatives," *IEEE Communications Magazine* Aug. 1995, pp. 34–46.
3. Bernstein, L. "Arrangements for dynamically identifying the assignment of a subscriber telephone loop connection at a serving terminal," U.S. Patent 5,355,405 (Oct. 11, 1994).
4. Bernstein, L. "Routing to intelligence," U.S. Patent 5,392,277 (Feb. 21, 1995).
5. Berkowitz, G. "Operations efficiencies of broadband access networks," *ISSLS '96 Proceedings* (Melbourne, Australia, Feb. 4–9), pp. 210–215.
6. Lawrence, V., Smithwick, L., Werner, J., and Zervos, N. "Broadband access to the home on copper," *Bell Labs Technical Journal* Summer 1996, pp. 100–113.
7. Bucholtz, C. "Keep it working—Broadband services place new demands on network management systems,"*Supplement to Telephony* March 10, 1997, p. 11.
8. Harper, M., Robson, S., and Combs, C. "Method and apparatus for provisioning a public switched telephone network," U.S. Patents 5,491,742 (Feb. 13, 1996) and 5,416,833 (May 16, 1995).
9. Verger, J. "Any service, port, time," *Telephony* May 5, 1997, pp. 38–40.
10. Butler, T., Hacker, L., Jadhav, S., Jessup, A., Keffer, R., Lamm, I., and Weber, C. "System for providing communications services in a telecommunications network," U.S. Patent 5,528,677 (June 18, 1996).
11. Cerf, V. G. Personal communication to authors. Copyright 1995, 1996. Permission is granted to reproduce freely provided credit is given to Vinton Cerf.
12. Hiatt, B. "Unchoking the INTERNET," *Lucent Magazine* April 1997, pp. 8–12.
13. Daugherty, T., DeBruier, D., Greenberg, D., Hodgon, D., and Murphy, D. "Method and apparatus for establishing connections in a communications access network," U.S. Patent 5,386,417 (Jan. 31, 1995).
14. Daugherty, T., DeBruier, D., Greenberg, D., Hodgon, D., and Murphy, D. "Communication access network routing," U.S. Patent 5,381,405 (Jan. 10, 1995).
15. Doshi, B., and Heffes, H. "Performance of an in-call buffer-window reservation/allocation scheme for long file transfers," *IEEE Journal on Selected Areas in Communications* **9:**7, 1013–1023 (Sept. 1991).
16. Markoff, J. "Market place," *The New York Times* June 10, 1994.

3

Network Architectures for the Future

Network architecture forms the main infrastructure to meet enterprise networking needs. Today's telephone networks treat switching, transmission, and operations systems as distinct disciplines. The usually respectable tendency toward caution in adapting to change would suggest that new technologies be mapped into this methodology. This is not possible. The deployment of TMN illustrates the problem; it was so slow in coming primarily because of its complexity and the inertia of legacy systems.[1] Distinctions between switching and transmission equipment disappear as network elements become software-based. Incorporating network management functions in the network elements can make operating a network easier.[2] Only a radical change of mind-set can hope to provide the necessary cost benefits and ease of use.

3.1. THE PURPOSE OF NETWORK ARCHITECTURE

The full service, multiaccess network requires an architecture that accommodates many modules, many topologies, and many media; one possibility is shown in Fig. 3.1. A shared access infrastructure featuring a connection control layer connects network services and management to the modules that provide intelligent network capability, such as call set-up, routing, and transport. Connection control provides the pathways for a combination of these modules to work together to perform traditional network functions. The modules can be at different sites or side by side. Network engineers can match module size to anticipated traffic and can handle traffic for more than just a wire center. All modules have equal access to network management functions and to the data embedded in the network elements. The modules may be accessed by different line speeds, technologies, or media. There is the advantage of engineering bandwidth allocation and service selection close to the customer or deep within the network. This forces architectural decisions that incorporate network management as a primary consideration, in contrast to today's network architectures where sharp distinctions are made between the network elements and the feature needs.

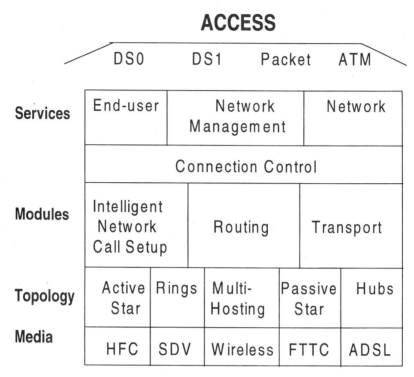

Figure 3.1. Architecture of network management for a full service network (multiservice, multimodule, multitopology, multimedia, multiaccess).

3.1.1. Changing Mind-Set

The change of architecture suggests a changed approach to thinking about operating networks. The switching, transmission, and OSS focus gives way to routing, transport, and network management. The new telephony networks will look more like the typical client/server computer network shown in Fig. 3.2. Clients connect to local servers in Local Area Networks (LANs).[3] Routers access remote servers through routers on the LAN that connect to the outside world through Wide Area Networks (WANs). The networks have the advantage of autodiscovery and automatic configuration management with in-band network management using SNMP. Network managers can "ping" components to detect problems but cannot easily isolate problems or remotely reconfigure their networks. These techniques supplement the telecommunications network management toolbox that concentrates on out-band monitoring and measurement. The new networks will withstand the types of outages experienced in client/server networks listed in Table 3.1.

3.1.2. Technology-Independent Architecture

The TINA-C model for telecommunications software is an architecture based on distributing computing concepts in a technology-independent way.[4] Magedanz

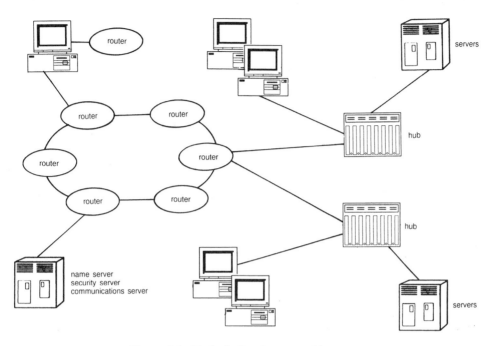

Figure 3.2. Typical client/server architecture.

Table 3.1. Outages that Can Propagate through Client/Server Networks

Broadcast storm	A broadcast is a special message or packet that all network hosts must receive and process. A broadcast storm is a condition in which broadcasts are overused, potentially completely disabling the network. Broadcast storms usually occur because of software errors
Babbling node	The transmission of random, meaningless packets onto the network; often caused by a failed LAN card
Runts	Packets that are smaller than the minimum length allowed by the network protocol, for example, 60 bytes for Ethernet
Jabbers	Packets that are larger than the maximum length allowed by the network protocol, for example, 1518 bytes for Ethernet
Router algorithm conflict	Routers use some variant of a shortest-path algorithm. If, for example, router A thinks that the shortest path to router C is through router B, but router B thinks the shortest path to router C is through router A, packets for router C will be sent back and forth between routers A and B. This usually occurs because of some breakdown in the router shortest-path updating strategy
Network paging	A (diskless) workstation runs a job that is too large for its memory and has to page over the network, causing very heavy network traffic

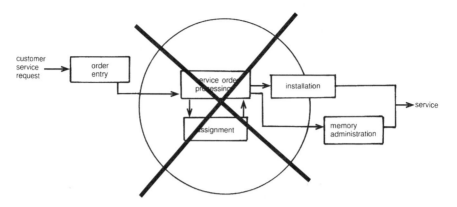

Figure 3.3. Reduction of effort triggered by dynamic provisioning.

reported that the current model of Intelligent Networks needs to evolve to an object-based model similar to that proposed in TINA-C. He suggests that the TMN model can itself be extended to call management and other telecommunication services.[5] His architecture provides a bridge from distributed processing networks to telecommunication networks and a faster deployment of TMN.

Including special feature client/server architectures in the self-identifying Network Interface Unit (NIU) can lead to $50 to $100 in savings[6] on annual operations costs for each residential line. The NIUs can determine available capacity from the network itself, associate dial tone at the edge of the network without maintaining separate databases, report status at all electronic points, poll for status throughout the outside plant network, and drop test for residual copper. Routing automatically eliminates loop assignment at installation, as the network dynamically assigns resources for local access. Local call routes are selected at call set-up just as in today's long-distance networks. Eliminating route dedication at the time of installation dramatically reduces provisioning and maintenance work because technicians have accurate and timely data. This leads to better inventory management that results in improved network planning, engineering, and service provisioning. With better information when customers call, there will be fewer incorrect dispatches, effective and efficient business offices, and happy customers.

Figure 3.3 shows the resulting changes in the provisioning process. The far more expensive applications for consumer broadband full-service networks such as telecommuting with two-way video at 384 kbs, instant video-on-demand, remote interactive classrooms, and telemedical treatment ride on the coattails of the operational savings. Without those savings, telephone companies will not be able to provide enhanced services economically. Users do not know the ultimate uses they may require, but once a direction is established, it stimulates a creative blossoming in that direction.

3.2. DYNAMIC SERVICES

Specialized ATM switches in COs will aid the unbundling of access networks. ATM switches can be placed in each CO to route customers' calls to the network servers they need and to route incoming calls to them based on tables in the ATM switch itself. This approach does not limit engineers to CO intelligent network services. It permits an orderly transition to the new architecture with a mixed network of old and new. The concepts of *routing to intelligence*[7] and *number portability* will preserve telephone numbers while allowing customers access to new services.

3.2.1. Routing to Intelligence

Routing to intelligence is the idea of placing databases at each access point so that the endpoint can reach any particular server that the application needs. Consider the example of e-mail: The signal goes to an ATM switch at the CO where the header information identifies the need, so that instead of going through a voice switch and then being sent to a packet switch, the appropriate routing is done using routing data in the ATM database.

All-purpose switches being designed by Lucent Technologies and WorldCom have integrated these routing capabilities. The network management problem is to populate the databases at each access point. The short-term approach is to build off the existing 800 service where there are centralized databases, though this requires several hops to get at information within the network and then returning to set up the circuit switch. This short-term approach does not take advantage of rapid routing at the endpoints.

Whether the long-term approach of the all-purpose switch or the short-term approach of using the 800 service databases takes precedence depends on which technology evolves more quickly and which proves more reliable. Routing to intelligence, however, provides an inherently more reliable network because there are no single congestion points.

3.2.2. Number Portability

The telephone numbering plan remains intact as the router in the ATM switch gets the customer to the right connection. Solutions to the number portability problem that impact current OSS infrastructure, class functions, and the geographic nature of the CO are unsuitable. Centrally controlled databases in the call path for all local calls are an equally unsuitable alternative because they require complicated simultaneous flash cuts for network elements, OSSs, and technicians.

3.2.3. Engineering Growth

Engineering server growth for a region is a natural evolution from today's numbering plan and maintains local life-line services. ATM voice is essential if carriers are to make ATM pervasive and cheap enough to support, for example, home multimedia technologies.[8]

3.3. NEW DIRECTIONS IN TRANSMISSION

One fiber can handle voice, data, and video traffic at gigabits per second, whereas copper is limited to a few megabits per second. For this reason, fiber cables are replacing copper ones as the backbone of telecommunications.

3.3.1. Amplifying Fiber

Fiber cable, just like copper cable, contains many separate strands that are bundled together and encased in an outer sheath. At one end of each fiber, a laser-generated beam of light carries the desired data and shines the modulated light into the cable. At the receiving end, a photodetector picks up the beam and translates it into electrical impulses. Again just like copper, a long cable run can cause too much signal loss in the fiber and the solution is the same: electronic signal regenerators added periodically along the cable run. The current problem is that these regenerators require photons to be translated into electrons for amplification. Then the signal is modulated back into photons by another laser that shines it down the next segment of optical cable. These regeneration points are expensive to maintain and operate and they require uninterrupted power. Eventually optical amplifiers that do not require active electronic devices may replace the signal regenerators so that the entire optical cable run can be a passive device and still compensate for transmission losses.

3.3.2. Power Costs

The cost of translating photons to electrons and the need to power POTS service limit the use of fiber in the loop to places that have a lot of traffic. As the costs of the equipment go down, the fiber will reach farther into the outside plant. Until then, a good compromise that gains many of the operations advantages of a pure fiber solution is using coaxial cable from the fiber node to the residence. With fiber in place, SDH can be used to carry voice, data, and video. Three versions exist: SDH-Europe, SDH-Japan, and SDH/SONET for North America.[9]

The SDH/SONET overhead channels could be converted to ATM cells and mixed with all other ATM cells to use the bandwidth more efficiently. Special Permanent Virtual Circuits (PVC) could be used as control paths for network management data and commands. The term *turbo-trunk routing* was coined to describe this marriage of transport with routing. Which engineering choice will finally take precedence, ATM over SONET as it is now, or ATM cells encapsulating SONET commands to handle bursty network management data, is a matter not of fundamental functioning but rather one of engineering efficiency and legacy investment.

3.3.3. Technical Issues

The following technical issues must be resolved before using the new network architecture:

- Signaling, including call set-up, billing, mixed audio and video conference calling

- Numbering plan compatibility

- Emergency services (E911)

- Lifeline services

- Directory services distinguishing video, audio, and data customers

- An end-to-end architecture plan with a product road map showing 6-month snapshots

- Isochronous voice and video or asynchronous cell switching with no more than 40 to 60% engineered utilization

- Standard ATM cell sizes that treat the needs of satellite communications

- Standard ATM cells that match IP packet needs

3.3.4. Wavelength-Division Multiplexing

SONET provides an electronic standard for high bit rates, that is, a control of optical fibers that combines a method for converting from photons to electrons and a powerful set of network management capabilities. Wavelength-Division Multiplexing (WDM) is strictly a matter of multiplexing message signals on photons and provides an optical solution that increases available bandwidth. Instead of using only one laser, multiple lasers operate at different colors or wavelengths on the same fiber. WDM lets carriers realize end-to-end in-band testing without taking the customer out of service. Test signals are sent down the fiber at a wavelength different from the one used for the customer signal.

WDM lacks network management and performance monitoring capabilities at the optical layer, the most important factor that will delay wavelength networking. By combining WDM with SONET, a carrier can increase the capacity of its in-place transmission equipment without having to install new cables. Alternatively, adding more optical fiber in the form of Optical Add/Drop Multiplexers (OADMs) can be attractive. In any case, it appears that the work of bypassing SONET is not practical.[10]

3.4. BEYOND HYBRID FIBER COAX (HFC) AND SWITCHED DIGITAL VIDEO (SDV)

One experimental WDM architecture uses two fibers with a loop-back feature instead of one to simplify operations. There are several key features that a second fiber allowing an optical loop-back can accommodate:

- The optical health and characteristics of the fiber cable can be monitored and measured while the customer's optical-to-electronic translating equipment is off. This relieves the need for continuous communication between the carrier and customer equipment and the use of power provided by the customer. This feature supports the connectivity assurance needed for lifeline service while reducing the drain on carrier supplied power.

- The Optical Network Unit (ONU) located at the customer's site can "sleep" in an ultralow-power (submilliwatt) quiescent state when not in use. Reduced power consumption makes it possible to reduce the complexity of the electronics; this reduces equipment and maintenance costs. For example, a penlight battery backup could keep the ONU up for 2 months instead of the 8 hours planned for single-fiber WDM. This dramatically reduces battery costs and replacement.

- A small set of wavelengths can be both down and up channels, conserving optical spectrum by using dense WDM rather than spectral slicing.

- The carrier can reuse fiber at spectral locations other than those originally planned. This relaxes the serious engineering risk of miscalculating the degree of communications growth. Other architectures cannot do this, so they use channel spacing wide enough to accommodate the known temperature variations of glass. This uses too much spectrum and makes additional services difficult to add.

- The ONU can ignore its role in a WDM network because there is no special record-keeping required at the termination, a particular frequency having been selected at the time that the connection is made and downloaded to the ONU at the time of selection. Everything is broadcast down the line and detectors pick off what is meant for a particular application. When lasers become cheap enough to install in the home, carriers must still know their own serial number for the equipment that works at each home in single-fiber architectures.

This experimental approach suggests how network management features can be designed into the architecture of new network elements, in an environment of much higher bit rates and diverse service mixes.

3.5. ISDN DEPLOYMENT

Worldwide, industrialized nations are deploying ISDN. Trials of ISDN in several large U.S. companies have demonstrated that where the customer is sophisticated and the trials are extensive in services, customers have bought in enthusiastically. John Mayo, a past president of Bell Laboratories, described the two forms of ISDN as follows:

Basic Rate ISDN offers speeds of more than a hundred kilobits per second over existing residential copper access wire, and Primary Rate ISDN offers speeds of more than a megabit per second using specialized copper access plant. With new enabling technologies, however, Primary Rate ISDN can potentially be provided over existing residential copper wire as well. Basic Rate ISDN can bring conference-quality video, data transport with graphics displays, high-speed facsimile, and multimedia communications to a home, office or school. Primary Rate ISDN can go a step further and support several simultaneous users, or support large-screen video. With consensus among service providers and equipment vendors, today's hundreds of thousands of ISDN users can potentially be expanded to many millions within three years.[11]

The difficulty of provisioning ISDN service is limiting its growth. Using dynamic provisioning techniques that incorporate the TR303 or V5 Time/Slot Interchange (TSI) standards, provisioning of ISDN service becomes practical. With preventative maintenance techniques, a projected $30 to $70 ISDN provisioning cost is avoidable. The combination of the technology push and the market pull can combine to produce a richly productive infrastructure (see Fig. 3.4).

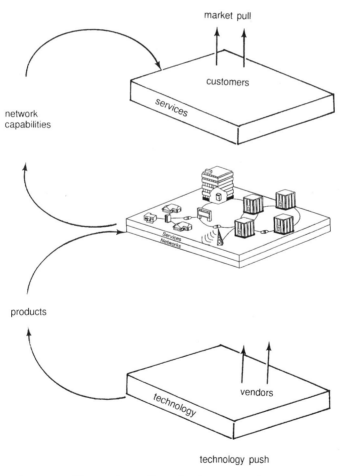

Figure 3.4. The market pulls technology to produce revenues.

3.6. TRIALS

Researchers, consumer video services, and network element providers have collaborated to create a comprehensive platform for the delivery of interactive video services to consumers in their homes via standard television. These services include movies-on-demand, interactive television, and video telephony. Experience with these demonstrates that the costs of such entertainment in terms of artists, information providers, service providers, and equipment manufacturers exceed the customer's willingness to pay. Additional uses beyond entertainment are needed to make this economically viable. While applications gain market acceptance, network providers can save substantial operating costs by using a hybrid fiber/coax network even if it is only to improve their telephony operations.

The Huamei Communications Company demonstrated China's first ATM network in Guagzhou, southern China. The trial network combines ATM-based switching with SDH transport and enhanced multimedia modems to connect advanced video, audio, and data communications directly to users' terminals. Users can integrate high-speed multimedia applications into their daily business operations without annoying communications interruptions or separate network connections. Guan Yi Zhang, assistant general manager of Huamei, sees this approach as "using the most advanced networking technology in the world to lay the foundation for similar developments across China."

3.7. BENEFITS OF THE NEW ARCHITECTURES

Profit, elegant design, reliability, and a supportive environment are benefits that accrue to full-service ATM-based architectures. On the other hand, there is an argument that can be made for special-purpose networks as being cheaper and more manageable.

3.7.1. ATM-Based Designs

The first benefit is profit. Customers will have access to network hubs where they may obtain video services and a broad array of other services, generating additional revenue. The second benefit is the simplification of the network. Much of the "connect the dots" provisioning problem vanishes.

The third benefit is enhanced reliability. Future indestructible networks will have two foci of robustness: the element level and the network level. Advanced superreliable switching elements prevent failures, detect problems, and automatically recover. Distributed architectures and disciplined development methods design reliability into the network elements. Fault-tolerant software is the foundation of the element designs. It constantly monitors element operation and it is defensive and self-healing. The impact of component failures is minimized by having automatic recovery, exhaustive redundancy, and generous sparing. The new network elements will have far fewer components to manage. The best can achieve a 60%

decrease in CO plug-ins and occupy 40% less space. These superreliable elements allow engineering at the network level, where great improvement in reliability can be envisioned.

At the network level, special routing designs are used to isolate elements when they are in trouble. These same algorithms are also used to control congestion. The routing algorithms monitor critical network resources and, while the network operates, find new paths when old ones are in trouble. The Internet is especially robust because each router is designed to link to at least two other routers and routing algorithms are distributed. The voice network relies on sensing trunk overflows as an indication of trouble and using preplanned routing around trouble spots. Care has to be taken that there is no instability in the routing algorithms, but otherwise this scheme ensures considerable robustness.

The fourth benefit is that customers will be able to define services and manage them in a supportive environment. Network operators will manage each desktop component of the customer's equipment separately, offering transparent client/server computing through network services. The powerful concept of self-provisioning lets customers buy network capacity as we now buy electricity. Customers plug in and are charged by the amount of power used, with no provisioning being necessary.

Achieving a self-aware, self-adapting, self-provisioning network as shown in Fig. 3.5 requires skillful planning. A 10-year evolution from today's networks assumes that ATM-based designs and today's circuit switching design coexist and that current numbering plans are supported. Additionally, feeder-by-feeder network upgrades will be possible so that flash cuts of COs or regions will not be required. Sprint Corporation leads the effort to replace circuit switching systems with full-service ATM-based networks.

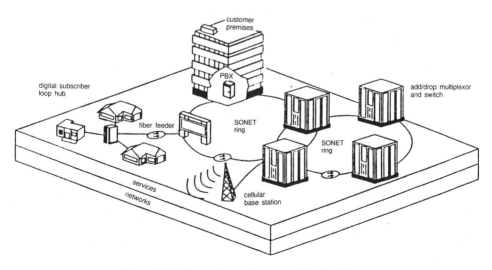

Figure 3.5. Networks evolve toward self-reliance.

3.7.2. The IP Alternative

There is no universal agreement that full-service networks are needed. Some say that special-purpose networks separating voice, data, and video will provide cheaper and more manageable solutions. Each network can be optimized for the traffic it carries. The inherent robustness of the Internet design provides reliability for these networks. The partitioning approach minimizes congestion problems by separating the different traffic. Almost immediately, however, designs have developed that hybridize IP with other technologies and dilute its benefits. The growth of gigabit IP switches, studies of placing IP directly on SONET networks, and changing the software in ATM switches so that they become IP switches are examples. The market excitement about voice over IP may be premature because the congestion and inherent delays in the voice communication are still being studied.

3.8. WIRELESS ISSUES

Personal Communications Services or Systems (PCS) support wireless access including cellular, cordless telephony, wireless data, and satellite-based services. PCS provide personalized voice, data, image, and video communications that can be accessed regardless of location, network, and time. Some writers use the term *nomadic* to mean mobility with plug-in capabilities.

3.8.1. What Is PCS?

The key defining feature of PCS is wireless mobility, whether for computers, cellular phones, pagers, or any other device. The key defining problem of PCS is the layout of antennas for radio frequency propagation between transmitters and receivers. The network management problem is the same as operating a wire network with an overlay of complexity added by the possibility of both sender and receiver being in motion during a transmission through various providers webs of antennas. The key goal of PCS is to maximize coverage and provide enough capability to meet customer needs while using the least possible bandwidth to avoid using up the radio frequency spectrum.

3.8.2. How Does It Work?

The general case is that a mobile terminal sends a radio frequency signal to an antenna at a base station. The base station is connected by wire to a mobile switching center, which is then connected to wire networks.[12] The antenna at the base station is a particularly vulnerable point. When it fails, all transmissions within its radius are lost. The mobile switching center is also more vulnerable than any CO. It handles more volume than a CO and does it with less reliability.

3.8.3. Vulnerabilities

The nonstandard, unregulated business arena in the United States has produced many suppliers of many services with little attention to network management constraints. The European market has had to work within more government standardization, which has initially produced greater reliability. The International Telecommunications Union in Switzerland is considering proposals for standards concerning third-generation cellular telephones (first generation was analog in 1983; second generation was digital) which will define many issues for wireless vendors.

Worldwide, the cellular business is growing rapidly. In the United States, the market pressure has been to get the antenna patterns correct and little investment has been made in, for example, partitioning and protecting the mobile switching centers to make them comparable in security and reliability to COs. Goodman has remarked that, "Because [the communication] characteristics of subscriber behavior are hard to predict, operators of personal communications systems rely heavily on network monitoring techniques and on planning strategies based on measurements and predictions. For the most part, these techniques are not addressed by the standards that govern other aspects of system operation. They are proprietary, competitive tools of each company."[13]

Mobility management is a significant network management problem. One must be able to locate a roaming mobile terminal to deliver calls and also maintain the connection while the roaming mobile terminal changes its point of attachment. Akyildiz *et al.* remark that location management protocols deal with querying and storing information in location databases and are therefore universal, but the algorithms for handoff protocols are network-protocol-dependent, which complexity argues strongly for the development of a standardized network architecture.[14] Different providers also bring with them problems of routing and billing arrangements.

3.8.4. The Role of the Internet

TCP/IP is the lingua franca of the Internet, and if its capabilities concerning data transport can be employed, wireless communication to the Internet or a private IP-based network can occur. Efficiencies can be employed such as the technique of *tunneling* described in Section 2.9.2, where an extra header makes it possible for the mobile laptop user to get to an Internet service provider without having to route all messages through his or her home host computer.[15] This eliminates the need to relay messages from place to place and sets up a direct path. The network management problem is to balance the traffic load in the networks, recognizing that mobile users are moving about, wireline networks are sensitive to congestion, and wireless networks are sensitive to noise. Sometimes proxy routers snoop on traffic at the boundary between wireless and wireline networks with different error recovery schemes because both managers must track the boundaries where problems are most likely to happen.

3.9. TECHNOLOGY IMPLICATIONS

Fault-tolerant software and object-oriented technology will be used within the network elements, for network and service management functions, and for presentation layer software. An asset base of objects will be the prime source for object libraries.

Some possible results of replacing today's physical connectivity might be mass storage server/router systems for distribution of products, and end equipment automatically identifying the customer so that free movement around the network is encouraged. As ATM costs go down with increased use, it will be necessary to have the appropriate tools to measure QoS in this area. The parameters to be measured would be cell loss rate, the cost of overheads implied by small ATM cells versus special service networks, some measure of latency of networks or applications, and PVC and Switched Virtual Circuit (SVC) costs.

Some caveats are also necessary. A single technician interface and a single system administration presentation layer must be available from all network elements with human factors experts responsible for a coherent interface. Also, analytical indicators of the "goodness" of these networks are needed. Tools must be developed to measure the costs of many individual factors such as each ATM cell, PVC, SVC, access trunk, Intelligent Network (IN) port, and circuit.

3.10. A VIEW FROM THE CABLE INDUSTRY

ATM can play an important role in the cable industry's efforts toward digitalization. Because ATM is built on fast-packet data transport technology, it can be used to maximize the functionality of a high-bandwidth medium. This allows a cable company to use existing coaxial distribution plant to the maximum while moving to a full-service network.

The cable companies see ATM as a way to bring a big piece of the multibillion-dollar markets of arcade video games and the electronic equivalent of the video store into the home. The cable industry is a major player in the development of high-capacity servers.

3.11. DISCONTINUOUS INNOVATION

Stephen Jay Gould, the eloquent evolutionary biologist and paleontologist, observes that most of the time nothing much happens, despite the validity of the theory of evolution as change through time. "Niles Eldredge and I have tried to resolve this paradox with our theory of punctuated equilibrium. We hold that most evolution is concentrated in events of speciation, the separation and splitting off of an isolate population from a persisting ancestral stock. . . . [C]hange occurs in infrequent bursts and stability is the usual nature of species and systems at any moment."[16]

As we noted at the opening of this chapter, change in the telecommunications industry has generally been a slow mapping of improvement onto a basically stable organization. The Internet changed all that. It was a dramatic mutation, what we would call a case of *discontinuous innovation*, that produced radical changes in profiles of use because of how easy it was to use this "sprung-from-the-head-of-Zeus" new service. Unfortunately, it arrived without adequate network management features. Companies now using the Internet must deal with congestion problems and security breaches as they rely on it for mission critical function. Today's freewheeling, send and pray Internet service is not good enough.

The future belongs to the service providers who invest in new network architectures. They will provide better services at lower cost and stimulate the growth of bandwidth-consuming applications. Today's successful companies will have trouble making the needed architectural leap because it requires discontinuous innovation. Continuous innovation is easier, but the new architectures call for significantly different ways of running the telephone and cable businesses.

The key is to invest in the future by providing today's services more efficiently and effectively. Savings mined from these efforts will pay for the network upgrades, which can then be used for broadband applications. With the more reliable embedded network management in these architectures, service providers can reinforce the public's perception of their quality service.[17]

REFERENCES

1. Glitho, R. H., and Hayes, S. "Telecommunications management network: Vision vs. reality," *IEEE Communications* **33:**3, pp. 47–52 (March 1995).
2. Bernstein, L. "Innovative technologies for preventing network outages," *AT&T Technical Journal* **72:**4 (July–Aug. 1993); "Arrangement for dynamically identifying the assignment of a subscriber telephone loop connection at a serving terminal," U.S. Patent 5,355,405 (Oct. 11, 1994).
3. Wood, A. "Predicting client/server availability," *IEEE Computer* **28:**4, 41–48 (April 1995).
4. Fuente, L. A. de la, *et al.* "Application of the TINA-C management architecture," *Integrated network management IV* (Sehti, A., Raynaud, Y., and Faure-Vincent, F., eds.) (Chapman and Hall, London, 1995), Chapter 37.
5. Magedanz, J. "Modeling IN-based service control capabilities as part of TMS-based service management," *Integrated network management IV* (Sehti, A., Raynaud, Y., and Faure-Vincent, F., eds.) (Chapman and Hall, London, 1995), Chapter 34.
6. Pugh, W., and Boyer, G. "Broadband access: Comparing alternatives," *IEEE Communications Magazine* Aug. 1995, pp. 34–46.
7. Bernstein, L. "Routing to intelligence," U.S. Patents 5,390,169 (Feb. 14, 1995) and 5,392,277 (Feb. 21, 1995).
8. Nolle, T. "Voice and ATM: Is anybody talking?" *Business Communications Review* **25:**6, 43–49 (June 1995).
9. Conard, J. W., ed. *Handbook of communications systems management*, 2d ed. (Auerbach Publishers, New York, 1991), p. 424.
10. Lutkowitz, M. "The advent of WDM and the all-optical network: A reality check," *Telecommunications* **32:**7, 29–31 (July 1998).
11. Conard, J. W., ed. *Handbook of communications systems management: 1993–94 yearbook* (Auerbach Publishers, New York, 1993).
12. Janakiram, V. K., Chandra, R., Kripalani, A. T., Rudrapatna, A. N., and Russell, J. E. "Network management needs for the wireless communication environment," *Journal of Network and Systems Management*, **2:**1, 7–27 (March 1994).
13. Goodman, D. J. *Wireless personal communications systems* (Addison–Wesley, Reading, MA, 1998), p. 60.

14. Akyildiz, I. F., McNair, J., Ho, J., Uzunalioglu, H., and Wang, W. "Mobility management in current and future communications networks," *IEEE Network* **12:**4, 39–40 (July/Aug. 1998). This entire special issue is devoted to PCS Network Management.

15. Kessler, G. C. "Future mobility," *Telephony* **235:**12, 48–52 (Sep. 21, 1998).

16. Gould, S. J. *Eight little piggies: Reflections in natural history* (Norton, New York, 1993), pp. 277–279.

17. Bernstein, L., and Yuhas, C. M. "Network architectures for the 21st century," *IEEE Communications* **34:**1, 24–28 (Jan. 1996).

4

Methods, Practices, and Diagnosis

Business practices are formulas followed throughout a telephone company by everyone from service representatives to plant managers to network administrators. They are elegantly refined activities, honed by millions of repetitions, which meet the needs of a process. A practice cannot be understood and certainly cannot be changed without an understanding of the process that generated it, because "Though this be madness, yet there is method in't." The whole process of selling a service, getting the equipment in place, turning on the service, and billing for it is called *service provisioning*. One element of service provisioning is management of the system configuration, including the equipment and the databases keeping track of the equipment and the customer information.

This chapter will answer four questions:

1. What happens when you make a phone call?

2. How does all that stuff get into place?

3. What happens when things break?

4. How will broadband change 1, 2, and 3?

4.1. WHAT HAPPENS WHEN YOU MAKE A PHONE CALL?

Figure 4.1 shows what happens when a customer at telephone number (TN) 250–1234 wants to go out for the evening and calls a theater at TN 731–2345. The explanation for what happens in this simplest of situations serves to introduce basic telephone terminology and concepts.

4.1.1. Basic Terms

The telephone in the home is connected to two copper wires, called a *pair*. This pair of wires is connected to other pairs to form a unique electrical path to one

49

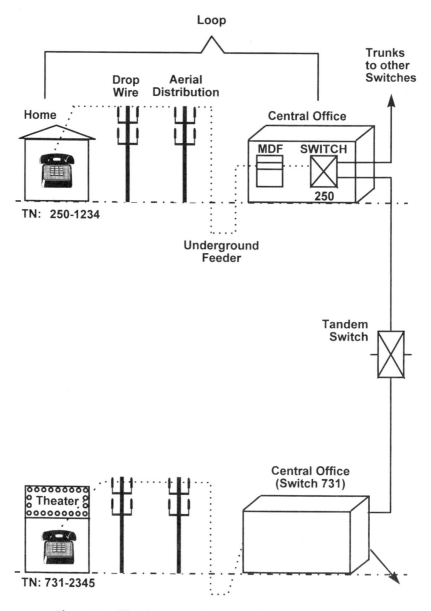

Figure 4.1. What happens when you make a telephone call?

of many COs owned by the telephone company. This assembly of wires is called the *loop*. The CO is where one or more switches are located; it is a major point of concentration for the individual customer lines. Inside the CO, there is a Main Distributing Frame (MDF) with unique positions that correspond to each customer's loop, telephone number, and the switching equipment used to originate dial tone. When all pairs are copper wires, the CO is the first part of the network where customers share telephone equipment. Figure 4.1 shows a CO with one switch for the exchange 250, the first three digits of the TN. The last four digits, 1234, point to the unique location on the MDF corresponding to that loop. All TNs beginning with 250 are connected to the same switch on the other side of the MDF, and trunks, which are large bundles of cables, link other switches in the same or other COs. The MDF is the boundary between Outside Plant (OP) and Inside Plant (IP). Everything from the customer to the CO is OP. Everything from the MDF within the CO is IP. Trunks leave the CO via trunk distributing frames, where they become part of OP again. The distributing frames are both a boundary and an isolation point where rearrangements are relatively easy to make.

Back to the customer wanting to see a movie; she lifts the handset off its cradle. Switch 250 detects a change of state in the loop—someone is off-hook. This happens quickly but not instantaneously because the switch must consecutively check all of the lines connected to it for state changes. As soon as the switch finds this change of state, it orders its dial tone generator to send the dial tone down the loop. This is a command to our caller to start dialing (a leftover verb from the days of rotary handsets). It is possible, though rare, for many people served by switch 250 to lift their handsets at almost the same time and perceive an annoying delay before hearing the dial tone. This is because the dial tone generator is shared among many loops and is engineered to a standard of less than a 3 second delay for the expected number of calls in the busiest hour of the day. Busy hours are determined by analyzing traffic patterns for the network for different classes of customers. Pre-Internet, busy hours for residences were 8 to 9 AM and 7 to 9 PM, and for businesses, 10 to 11 AM. As Internet use changes telephone usage, busy hours are changing radically.[1]

Our caller keys in the theater's number, 731–2345. She can hear tones as she dials. Each digit has unique frequencies (a "touch-tone") that the switch detects and translates into the appropriate digits. The switch in the CO examines the TN 731–2345. A switch is a collection of equipment that connects loops to loops or loops to the rest of the telephone network. When the switch determines that the call is not for itself, that is, it is not a 250 call for one of the other loops it is serving, it sends the call on its way to another switch. A signaling network finds an available path from switch 250 to switch 731. This process is called *call set-up*.

Once the path to switch 731 is set up, switch 250 sends the call out on a trunk. Trunks usually operate on four-wire combinations so that the electrical characteristics of the connections can be carefully controlled. Trunks connect switches to switches. In Fig. 4.1, the call is routed through an intermediate switch, called a *tandem switch*. Tandem switches connect trunks to trunks. In copper networks, the fundamental network management difference between loops and trunks is that the loop route is selected when service is installed at the customer premises and the trunk route is set up when the call is made. Call-by-call routing is used for detecting and avoiding congestion in the trunk network.[2]

The call follows a route selected by the signaling network through the tandem switch to switch 731. This switch recognizes the call as "self" and uses the last four digits to send a signal down the loop to ring the phone at the theater. If the theater's phone had been busy, the signaling network would have detected that state and sent a message to switch 250, which would have ordered the busy signal generator to send a busy signal to the customer on its loop 1234. If call set-up were not successful, a different, faster-tempo busy signal would have been sent back to the caller. In this case, our caller gets the movie selections and times and the telephone company earns another coin for its shareholders. From this simple structure, enormous complexity and variety in network construction has grown and volumes have been written to describe it. For this discussion, let us go no deeper, but refer to Appendix B for a listing of selected works that elaborate more fully on the layers of detail.

4.1.2. What Alters This Picture?

Modern switching equipment uses electronic TSI techniques following the TR303[3] or V5[4] standards to move MDFs from the CO into the field, closer to the customer. Digital Subscriber Line (DSL) equipment assigns a time slot to each loop and models the time slot as a virtual cable and pair. The effect of this is to eliminate the need for technicians to physically move connection wires at the CO. Many DSL frames in the OP, each handling a few feeder or feeder plus distribution cables, move the rearrangement activity closer to the endpoint and therefore the complexity at any one DSL frame is reduced. The degree of broadband readiness is a function of how close the link to the broadband network is to the customer. Less complexity means less work is done by people and more by computers. In a totally broadband network, this link to the broadband network would be at the customer's premises and connections would be set up automatically by the Core Access Networks.

By design, the architecture of the HFC system includes the following features to facilitate automatic provisioning without craft dispatch:

- TR303 or V5 interface with the Local Digital Switch (LDS)
- Self-identifying network elements and components
- Placement of an NIU on the customer's premises

The TR303 or V5 interface provides a dynamic interface to the switch that allows concentration to be distributed between the switch and the access system. Even if the overall level of concentration is kept the same as in traditional copper access systems, significant savings can be made in both hardware and operations expenses. The TR303 or V5 interface makes use of a Call Reference Value (CRV) that allows a simple association of the switch line equipment to the logical distribution facility. The logical facilities need only be associated with the physical port at the NIU. This association is all that a Host Digital Terminal (HDT) needs to store for call processing. The switch specifies the CRV for the terminating call in the HDT and the particular link between the HDT and the switch. The HDT then dynamically sets up a path to the physical NIU port that maps to the CRV and places it on the specified path to the switch. Originating calls are handled in a similar fashion with the HDT notifying the switch which CRV is placing the call.

Placing the NIU at the customer simplifies both provisioning and maintenance. The physical ports serving the customer premises are explicitly known and cannot be rearranged by craft without appropriate messages being sent to the HDT and OS. This keeps the records accurate and also gives specific knowledge of the size and service abilities. The intelligent NIU provides full diagnostic, test, and alarm capabilities to the point of separation between the network and the customer. Faults are repaired before the customer can notice them.[5]

4.1.3. Usage Patterns

Another factor altering the basic nature of the telephone call is that the Internet has changed the whole landscape of usage. Average call holding times of 3 to 4 minutes are breaking down. Busy hours are unpredictable, changing with the news of the day and the local usage profiles. Equipment engineering becomes a nightmare of unique tailoring to each CO.

On the brighter side, with the introduction of broadband networks to the customer, there is an opportunity to have call routes selected through the OP when the call is made rather than when service is established. This lets the telephone company share equipment in the OP network and eliminates the need to dedicate wires and equipment to a customer even when the customer is not using it. The most significant economic advantage is that the need to keep track of all of the connections in the OP for each customer is eliminated. This eliminates expensive databases, simplifies the process of setting up customer service, eases maintenance, and reduces the need for very expensive installer visits.

The basis for all of these advantages is the simple observation that even though the same route may be chosen for every call, the act of choosing it eliminates the need for tracking connections. This allows trunk administration to be part of the OP and eliminates entirely the need for loops and their administration.

4.2. HOW DOES ALL THAT STUFF GET INTO PLACE?

Service provisioning was originally entirely manual. From Mr. Bell until fairly recently, somebody had to touch every step along the way to make the phone service work. The objective of automated service provisioning has been to totally remove human activity from the process (*flow-through*) and concentrate human intervention in those areas where computers cannot do the total assignment. Repair systems use the concept of flow-through to mean the mechanized acceptance of the customer complaint through automatic testing, analysis and trouble clearing or automatic dispatch. The ideals of flow-through and instant service are only partially obtained. Sometimes the computer finds that it cannot locate the customer address in its records or can assign only one part of the plant, inside or outside. For some Service Orders (SOs), the computer will not be able to efficiently assign anything and one must return to a manual process. Let's examine that process.

4.2.1. Service Order

Everything that happens in a telephone company starts with a customer request for service. The service representative writes an SO that captures what the customer wants, where he wants it, when he will get it, and how much it will cost. The SO drives the process for all of the various functions needed to make a telephone work. It is used to select equipment for the customer, to provide instructions to telephone craft, and to modify other computer systems that handle billing, repair, and directory compilation. That telephone companies now use a computer-based Service Order Processor (SOP) to create and manage the order as it flows through many departments does not mean that the process is automated. Human intervention is mandatory; therefore, it is still a manual system that is being described here.

Information on an SO is coded in Universal Service Order Code (USOC),[6] a telephone company language. There are five main types of SOs:

- IN or I order: indicates that service is to be installed at a location.

- OUT or O order: indicates that service is to be removed.

- FROM AND TO or F&T order: indicates that service is to be taken from one location and moved to another. From the service view, it is the same as an O and I order except that the customer's existing billing records are retained and only the address changes.

- CHANGE or C order: indicates that the customer is staying in place but wants some change made to the service. Examples include adding call waiting or central office answering service.

- RECORD or R order: indicates that some information in the customer record will change but that no change in service is needed. Over half of the SOs generated by a typical telephone company are of this type.

4.2.2. Manual Service Provisioning

Figure 4.2 shows the process of manual service provisioning. Activities occur across six blocks of time. First, the customer requests service. The service representative fills out a contact memo during the customer conversation. During the second time period, the contact memo is given to a clerk to key the SO information into a computer running the SOP software. In most cases, these two steps have been compressed through the use of minimal input systems that tie existing records to the SOP, so the service representatives need enter only the current request directly into the SOP. For a new customer, all information would need to be added because there would be no preexisting record.

In the third time period, the SO is printed out at an assignment center and a clerk chooses the outside and inside plant equipment. The records containing the possible choices can be paper or computerized files.

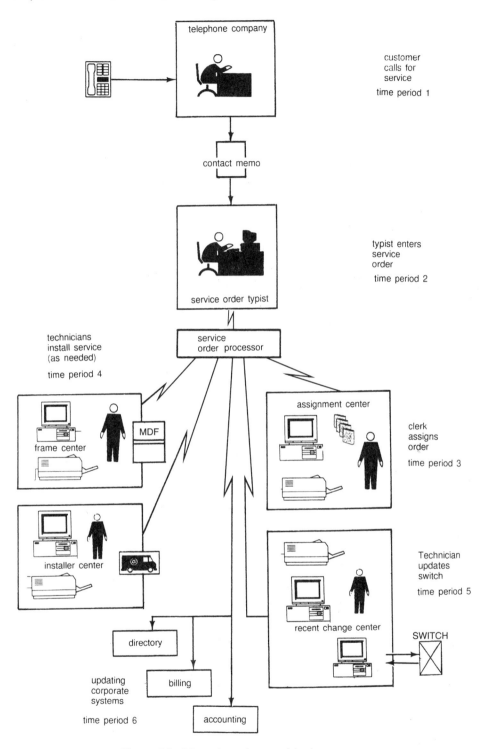

Figure 4.2. Manual service provisioning process.

During the fourth time period, the SO with the assignments on it alerts technicians to install the service by doing work in either the CO or the OP as needed. The frame center receives the dispatches for all of the COs in a particular district. The installation center handles the dispatches for all of the OP work. The goal of the assignment clerk is to reuse as much as possible of the plant already in place to reduce the need to dispatch these technicians.

After whatever physical work was needed is accomplished, technicians in the Recent Change Centers update the memory of the computer-based switches to activate the service in the fifth time period.[7]

Finally, in the sixth time period, corporate systems such as directory, billing, and accounting are updated. The databases supporting the repair systems are also updated and the assignment databases change the status of the pair from pending to working.

In summary, the functions of a service provisioning system, whether manual or automatic, are as follows:

- Accept, edit, and process SOs

- Interpret the SO; translate it to a request for "so many pairs at such-and-such an address"

- Assign the pairs; find unused pairs for the service or rearrange existing pairs

- Assign CO equipment; connect the assigned pairs to the MDF for the selected TN

- Request that craft make the needed connections if none already exist

- Activate dial tone

- Rearrange OP as new pairs are installed in anticipation of neighborhood growth

- Assign new pairs when repair is needed

4.2.3. Complications

If there were only one type of service, if there were only one piece of equipment to do each job, if all computer systems were compatible, if people could actually remember where they live and recite that accurately to the service representative, and if natural disasters never happened, this provisioning thing would be a snap. But that's not how it is. For example, it may seem a small matter to expect the customer to know where he would like his new service to exist, but fully 40% of the time people do not know their address well enough for an installer to find it.[8] Telephone companies have tried many techniques to solve this problem. The service representative can often use a computer-based street address guide to work with the customer to find the correct location. With broadband networks, the opportunity for automatic discovery exists because the equipment left in place at the customer location can describe where it is sitting. If the customer calls the business office from the location where he wants the new service, the network itself solves the elusive address problem.

Let's take one small but awesomely ugly detail as an example of myriad recordkeeping details demanded by this process. Until the 1980s, all records of how the OP was used were kept on paper forms called Exchange Cable Customer Records (ECCRs). This smudged, erased, written-over record kept track of all of the installed equipment, the connections between the various pieces of equipment, and the status of all equipment. These could be the serving terminal connecting the drop wire from the house to the rest of the telephone network or any of the many wires used to make the connections. Every change meant another notation on the ECCR. Cables contain many pairs, usually grouped in 25- or 50-pair packages and color coded for ease of identification. Each pair is twisted together to minimize the phenomenon of *cross talk*, which means being able to overhear conversation from one pair to another pair. Good-quality wires, attention to adequate signal strength, and twisting the wires eliminate this problem for POTS calls. Midband and broadband connections, however, spill more easily from pair to pair, so special engineering is required to keep special service circuits physically separated.

Manufacturers of the equipment that is used to put high-bandwidth services on copper pairs specify the separations and the assignment system or engineers make sure that these rules are followed when service is provided. The special considerations when copper loops are used to provide midband service such ADSL were discussed in Chapter 2. The three flaws in the ADSL/VDSL vision are:

- Expensive loop qualification

- Process for tracking "left-in" ADSL equipment is missing

- The continuing need to analyze plant changes that could violate ADSL design rules

4.2.4. ECCRs Disappear with Broadband

In a broadband network, the ECCRs and the computerized filing systems that replaced them all disappear. The special service engineering tasks disappear. The initial connection at the customer's premises is the gateway to the broadband network and only that termination point must be remembered.

Long distance and local-exchange carriers also face installation and maintenance problems. A study of DS-1 service several years ago[9] showed a 40% annual growth rate. Monthly customer trouble report rates were 3 to 4 per 100 circuits. The provider did not have automatic test equipment and the biggest problem was getting customers to release a circuit so that they could wire-in monitor points. Rather than continue to create service outages for the customer, they decided to install smartjacks for new services to allow them to run loop-back tests. This investment paid off in happy customers, preventive maintenance, and reduced trouble rates. Fewer introduced errors were a serendipitous by-product. Because there is a high correlation between monitored performance "hits" on broadband lines and customer complaints, good practice dictates investment in test gear and network technology that can be remotely configured.

4.2.5. Status Issues

Another major activity in provisioning the copper network is trying to keep knowledge current about the actual status of a particular service. When a loop is first assigned but before service is turned on, the status is *pending-in*. The reverse, when the customer asks that the service be disconnected but it is not yet deactivated, is *pending-out*. The systems used to administer the OP must know about all of the wires and equipment, where they are, what the possible connections are, and what the actual connections are. To complicate things further, they must know what the plan is for the future status of the loop. If a residence were vacant, the loop would be left in place waiting for the next occupant and the loop status would be *dedicated*. When the service is turned on, the status changes to *working*. If the loop has a problem, it is taken out of service, referred to the repair work center, and marked as *broken*. When a pair is not associated with a residence, its status is *spare*. And this is only the tip of the iceberg when it comes to tracking status. In a broadband network, where it is necessary to retain only the termination point of the service, the matter of status becomes trivially easy. The burden is entirely on the customer to notify a computer to change a bit, instantly, to turn service on or off. Everything else is irrelevant when the call path is selected with each call. Broadband networks embed equipment identifiers, respond to polled inquiries, and note customer data. With this information, equipment manufacturers can incorporate the ability to self-assign, self-test, and activate service on demand. These features are the basis for dynamic provisioning as defined in Chapter 2.

4.2.6. Timeliness Issues

Telephone companies are measured by their response to meeting customer due dates. When they cannot meet a due date because loops are unavailable, the order is *held*. When orders are in held status beyond a set number of days, they are sent to the OP engineering centers where the physical inventory is kept and a frantic search ensues for loops. Often there are still paper blueprints, although many telephone companies have used Computer-Aided Design (CAD) systems to store some, if not all, physical plans in a computer. These physical views of the OP are the basis for engineering incremental changes to the plant to accommodate the need for growth in specific areas. Once new cables are installed, they are made available for assignment through the use of cable throws.

Cable throw practices direct the rearrangement of the plant to make best use of existing cables in concert with new cables being installed. Working service is sometimes moved from one loop to another to balance the use of the OP. When this happens, the cable and pair assignment to a customer changes. The databases tracking the cables and pairs must also change.[10] During the process of changing the loop, service may be available on both the old and new loops at the same time. To bridge a customer's service on two loops for the purpose of minimizing the service outage time during an engineering-driven transition, a *half-tap* on the CO MDF is used. The half-tap must be removed from the formerly working loop before it can be used as a spare loop. Special half-tap management features have been included in assignment systems to schedule orderly removal of half-taps for loops becoming spare, but to permit earlier removal of specific half-taps on loops needed for customer service.

A loop may be attached to several residences. Once one residence is selected, the other connection points are left open. These places are called *bridged taps*. When loops are used for anything other than POTS service, their electrical characteristics must be checked to make sure that the service will work. These are called *special service loops* and their SOs are sent to an OP engineering center to evaluate the design of the loop. The length of the loop, its resistance, the presence of back-taps or half-taps, and the kind and quality of all equipment in the loop makeup are considered for their effect on service quality. Ten to fifteen percent of all loops are used for special services and, of these, 3 to 5% need to be upgraded from two-wire to four-wire design to handle the service. A few services need even more wires to work. As loops are used for midband services, these situations will occur more frequently. With the cost of qualifying a loop being high because the process is time consuming and requires skilled people, costs to provision telephone service will increase three to five times[11] unless steps are taken to simplify or eliminate the need for loop engineering. Of course, the problem is moot in broadband because the network searches for the correct electrical characteristics for each unique call path.

4.2.7. Broadband Eliminates Assignment and Installation

Although it is often difficult to get telephone service installed, once it is working, it works remarkably well. Assignment practices try to minimize the number of rearrangements to the physical plant by reusing connections in place. Elaborate schemes to dedicate loops to customer locations are used throughout the telephone industry. They all have one goal in common, namely, to keep the amount of craftwork required to a minimum. Any method or practice that reduces the need to dispatch craft saves extraordinary amounts of money. These approaches have had some success, but are not perfect because of mismatches between the recorded data and the actual way the plant is being used. In addition, the desire of each work center to increase productivity sometimes at the expense of attention to recorded details, the introduction of new and complex services, and human error break down dedicated plant practices.

Broadband offers the possibility of eliminating the entire assignment and installation process by embedding the establishment of the call path within the call set-up capability of the network itself. The breakthrough idea[12] is that even if the same path is selected every time a call is made, the ability to dynamically select the path eliminates the need for assignment and installation. Once a construction crew installs a loop to a customer's location, all that remains in provisioning to activate service is to open a connection via a computer bit change. Figure 4.3 shows that par-

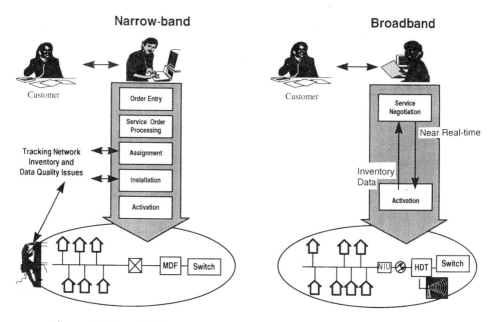

Figure 4.3. Provisioning paradigm shift with the introduction of broadband.

adigm shift. Dynamically selecting the path for each call eliminates assignment and installation, thereby removing an opportunity for databases to be changed, perhaps erroneously. With broadband, the databases are untouched, and the integrity is maintained. Also, the provisioning can be done in near real time rather than in a matter of hours or days.

Figure 4.4 shows the paradigm shift that occurs for maintenance with the introduction of broadband. The possibility for flow-through exists in repair once the network is self-monitoring and self-testing. When the network can sense trouble conditions arising, the repair crew can be automatically dispatched before the customer detects any negative effect in service. With such proactive maintenance, the customer is no longer the primary alarm generator; if the customer needs to make a trouble call at all, it will correlate to and corroborate reports already self-generated by the network.

4.2.8. Database Issues

The difference between manual and automatic provisioning systems is that the latter eliminates human intervention except for problem SOs. Service is activated automatically after the SO is entered into the system. Automatic provisioning systems can work with multiple databases or with a single common database. Work groups need data organized to suit different requirements. Some need a certain subset of the data available quickly and others need different subsets but fast

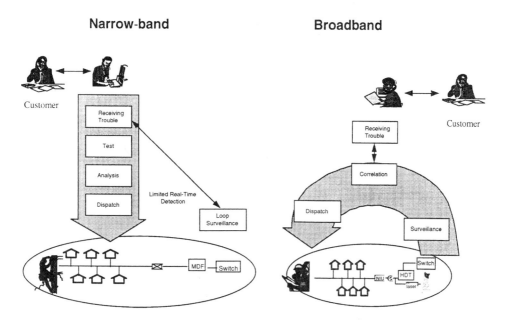

Figure 4.4. Maintenance paradigm shift with introduction of broadband.

response time is irrelevant. The assignment database design is based on the paper ECCR records. These paper records gave the assignment clerks easy access to the serving terminals, the cable pairs, and the cross-boxes. The complexity of the OP topology led to the invention of a hyperentity database system in the early 1980s.[13] This database design was able to handle the many ways the OP could be connected together. By 1990, relational database technology was robust enough to handle the data schema and transaction reliability needed for assignment functions.

Many of the databases grew with the systems to automate individual work centers. By the early 1970s, system architects realized that the available technology was incapable of creating one single system for automating the operation of the telephone company. As separate systems grew to provide machine aids to help improve local productivity, the need to link the systems to automate complete processes became apparent. Linking these distributed systems was accomplished at the cost of having duplicate redundant databases. New work centers were created to keep the databases consistent and accurate. Growing systems tried to handle new data types, larger numbers of data items reflecting new plant equipment and larger geographic areas, and growth in transaction load reflecting an increase in loops. But what works in a 20,000- to 30,000-loop CO often does not scale well to huge 50,000- to 100,000-loop COs. Midband services complicate this problem because there is simply more equipment to track. The SOP has a temporal database containing the TN, SO number, class of service, customer name and customer address, all information needed by both narrow-band or broadband services. In broadband networks, however, the database shrinks to a size comparable to those for interoffice connections and long-distance records. The existing architecture is sufficient to handle this reduced number of data items.

Distributed databases are a problem to maintain in a consistent and accurate state. Consistent databases tell the *same* story, whether it is a true or false story. Accurate databases tell the *true* state of real things. Database replication is the synchronized duplication of all or part of a database so that copies of the database can be distributed across a network for faster access. Replication allows dispersed users to share the same data in almost real time. It also allows backup to go on by using a replica while users continue to access the original. Unfortunately, the replicas begin to diverge immediately after replication. If all versions must be identical, some kind of transaction processing that updates them all is required. Database-to-database and database-to-physical plant reconciliation is expensive. When data are embedded in the network, the network equipment takes care of both issues because a network management system has the ability to do autodiscovery. This is possible in broadband systems.

4.3. NOW WHAT'S GONE WRONG?

Broadband might put assignment and installation out of business, but repair is eternal. When network performance departs from its design (to understate the customer hysteria usually accompanying this event), network diagnosis becomes necessary to get that hard-won service connection back. The network manager must find the anomaly, isolate it, take remedial action, and coordinate efforts to fix it. The good news with broadband is that the design that requires intelligent network elements also provides the opportunity to merge surveillance and testing functions.

4.3.1. Repair

The network manager, whether human or computer system, tries to achieve an economic balance among the costs of highly reliable network elements, preventive maintenance, and repair. The maintenance policies best for narrow-band networks are not best for broadband. The faster the network, the more the economic balance shifts from preventive and corrective maintenance to built-in component reliability, self-testing, and self-alarming capabilities. Network elements work faster, carry more traffic, and become laden with electronic and photonic components in a broadband network, so money is wisely spent on buying the best performing model of every element. This shift means that the human who could keep up with a copper network is overwhelmed by the speed and enormity of a broadband network; therefore, automatic OSS flow-through becomes desirable (see Fig. 4.4). Trouble diagnosis is indirect in a broadband network because of the nature of digital performance (recall the OK, OK, OK, DEAD syndrome). The network manager must rely on observing analog mechanisms to detect changes in network states rather than direct measurement as in a copper network.

4.3.2. Diagnosis

The concepts of *surveillance, testing, monitoring,* and *measurement* are seen, paradoxically, as discrete but often interchangeable activities, a mistaken impression that affects the way diagnosis functions. These terms define a continuum of activities used to keep a network operating. Surveillance and testing are both methods of access to information. Surveillance is the unobtrusive bridging or "camping-on" of equipment onto circuits without affecting either performance or user service. Surveillance can be continuous or on a scan/sample basis, occurring only when a result deviates from a norm. Surveillance has a network operation orientation. By contrast, testing is disruptive; a circuit must be taken out of service and reconfigured to apply a reference signal and a detector. Testing has a service orientation. Monitoring and measurement denote the use of acquired information once surveillance or testing access has been gained. Monitoring presents discrete events that have been recognized as unusual, such as failures or signals crossing preset thresholds.[14] Measurement is the selective collection over time of data that are later analyzed.[15] Figure 4.5 shows the relationship of these four diagnostic methods.

The data collected in this way must get from the equipment being observed to the network manager. *In-band management* is when the same facilities are used for the alarm messages and the customer circuits. Computers prefer this method, using SNMP protocols. When special circuits are dedicated to carrying the alarms and alerts, it is called *out-of-band management*. Telephony networks favor this because of a reluctance to rely on the network to manage itself. In broadband, both ideas come together because both are needed. Out-of-band management is needed

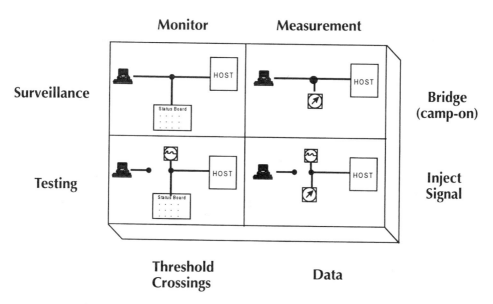

Figure 4.5. Possible combinations of network diagnostic methods.

to monitor the equipment when catastrophic failure occurs; SONET rings are a way of ensuring no single failure would cause a loss of data and control. In-band management is needed to reliably do equipment identification, and customer circuits with thresholds to reduce the number of false alarms are much more reliable than monitoring circuits.

While surveillance and testing grew independently in narrow-band networks, broadband allows integration of these techniques. The inherent flaws in network management systems that separate these techniques are becoming more apparent as broadband networks evolve. Fiber-optic systems allow many more customers to be served by a single facility than copper ones. The consequences of a single network element problem are more severe. Preventive maintenance is mandatory because the alternative of merely reacting to unforeseen outages that take large numbers of users out of service is unacceptable. This has led to the introduction of automatic failure detection systems that use preplanned reroutes to restore service as the troubled network element reports its condition.

4.3.3. Alarm Suppression

Surveillance can be continuous or sampled, and may alert, alarm, or both. Faults are reported to network OSSs, which correlate the alarms and alerts to locate the source. Then the OSSs trigger tests to pinpoint the exact cause and location of the faults. This technique worked well in narrow-band networks, but with the advent of broadband services, major network element failure is recognized at many points in the network, either explicitly or implicitly. This flood of data saturated the older OSSs to the point where they became useless when a serious problem occurred. This phenomenon is called *alarm avalanche*[16] and drove the invention of alarm suppression techniques.

The surveillance OSS keeps a hierarchy of network elements in its database. Alarms are processed according to priority based on the capacity of the network facility reporting the problem. Alarms from those that carry the most traffic are processed first and lesser-priority alarms are dropped if there is no time to process them before a new cycle of alarms appears. When the failing network element is identified, the OSS links all other alarms to it by referencing its database hierarchy. For example, if a DS-3 facility containing 28 DS-1 circuits were to fail, its surveillance equipment would generate an alarm, as would all of the DS-1 terminal equipment. To make matters worse, all of the switches connected to the DS-1s would also generate alarm messages. The surveillance OSS using alarm suppression would ignore all of the messages except for the one coming from the DS-3 facility. Then it would discover which DS-3 facility had a problem and would suppress all future alarms coming from the DS-1s associated with it. On the next cycle of processing, alarms from any DS-1 not associated with the DS-3 in trouble would be recognized. This delays the recognition of these alarms but has the advantage of not overwhelming the correlation ability of the OSS. Before these techniques were available, network managers resorted to baroque methods such as disconnecting the OSS from the alarm reporting circuits and then reconnecting momentarily to prevent the system from becoming overloaded.

Bouloutas *et al.*[17] show how fault identification can be distributed to reduce the alarm traffic in networks. They also provide a detailed analysis of how alarms are generated and then flow from the network facilities to the OSS.

4.3.4. Degrees of Trouble and Diagnostic Steps

Trouble can be of varying degrees. A *fault* is an abnormal condition that may degrade the network if it multiplies or spreads. An *error* is any deviation from specification. *Failure* is an error that impacts service.[18] The steps needed to diagnose trouble in any network include the following:

- *Trouble detection*: recognize fault, error, or failure in the network. The detection may be made by a customer, the network manager, a network element with built-in detection equipment or by an OSS.

- *Trouble notification*: alert technicians to the existence and severity of the fault, error, or failure. Once trouble is detected and the data describing it are available, technicians or other OSSs are notified so that they may correct it. The trouble notification may take the form of an indicator on a display, a trouble report, and/or an audible alarm.

- *Trouble verification*: check to make sure that trouble still exists and gather baseline data. Because some time may elapse between the onset of the trouble and corrective action, a check must be made that the condition still exists. Experience shows that many trouble indicators are transient. First priority is given to correcting verified troubles. In many network management systems, transient trouble reports are routinely processed to look for trends.

- *Trouble location*: pinpoint a trouble to a specific network equipment or facility. When possible, the technician or OSS will cause the network to route around the trouble. This is often the most difficult and time-consuming step, especially when the trouble is intermittent.

- *Trouble repair*: technician replaces the defective network element with a spare. An important but often overlooked part of this step is to return the network to its original configuration.

- *Service verification*: technician or OSS tests to make sure that the trouble is cleared.[19]

4.3.5. External versus Internal Monitoring

These steps are the same for narrow-band and broadband networks. The difference lies in the source of the trouble information. Narrow-band relies on *external* monitoring equipment. Telemetry and sensors detect facility and switch failures and send digital data or telemetry signals from the trouble source to analysis points.

Analog circuits are tested for troubles with noise, attenuation, and matching the electrical characteristics to the user's service.

Telemetry circuits are far less reliable than the networks they monitor and they cannot share network facilities with other traffic. Also, the telemetry signals must be translated to digital data to be usable by an OSS. In early narrow-band systems, the OSS did the translation. Digital loop carrier alarms were monitored to detect problems in narrow-band loops and special satellites were added to capture the alarms.[20] Gradually this function moved to be colocated with the network elements.

Because telemetry signals cannot keep up with broadband speeds, broadband networks rely exclusively on *digital data generated at the network elements* to detect troubles. Broadband networks often use channels multiplexed on the call-carrying facility for such data.[21]

By the mid-1990s, broadband digital facilities monitored bit error rates, but to no useful purpose in detecting problems. Bit error rates are useful during installation to determine a level of service; they cannot show any gradual degradation of the facility. Digital networks either work or they don't.

A key question is how often the circuit should be sampled to provide signals that can be used to predict a network failure. Surveillance systems are charged with detecting faults in network elements. When a network element deviates from its nominal performance, it may generate a signal or another piece of equipment may detect the 'out of specification' condition and generate a signal. Often spurious, non-critical conditions may occur, so many surveillance systems process fault indicators close to the network element. The threshold may be tripped after a specified number of faults occur or when a number of faults occur in a specific time period. For example, broadband networks often measure bit errors per second and establish thresholds for the QoS of the facility. Only when the number of errors exceeds the threshold in a specified interval is a surveillance signal generated.

4.3.6. Digital Network Performance Study

A study of long-distance networks in the United States and Japan[22] examined the in-service performance and transmission performance of digital network services. The study showed that bit error occurrences vary with time and tend not to reappear during out-of-service testing. In-service surveillance is essential for deterring actual bit error rates. The study went on to show that long-term bit error rates are inadequate to evaluate the performance of data or facsimile services. These results extrapolate to video services. Second or subsecond sampling is needed for in-service monitoring of broadband services. Even with this rapid sampling, broadband facilities fail too quickly for remedial action by human operators.

A secondary goal of the study, which was to create a mathematical noise model for broadband facilities for use by network designers, was not accomplished. This model still needs to be created.

Network managers monitor the performance of analog components of digital facilities to detect trends that may indicate gradual degradation. For example, the gain settings on lasers are sampled and transmitted to an OSS for analysis.

4.3.7. Case History

We have had personal experience with one mysterious trouble that took both machine and human talent to unravel. An expert system that processed sampled data found that even though the gain settings were all within tolerance at one facility, there were unusual periods of high gain followed by periods of low gain within 24 hours on winter days.

The network manager asked the field technicians to inspect the facility and it tested OK. When they reported "no trouble found," the manager asked them to walk the facility route. As they approached a bridge across a stream, they saw the cable vault was split.

Water was leaking into the cable by day and freezing at night. Ice is a formidable barrier to laser light. When ice formed each night, the laser amplifier gain cranked up very high, but then it settled to nominal levels when the ice melted during the day.

This tale has a happy ending because the field technicians were able to splice in a new section of cable, only one frame of broadband data was lost, and customers stayed in service.

4.3.8. Testing in Broadband

The whole issue of testing changes with broadband. Historically, mechanized loop testing using centralized test sets became part of the repair center practices in the 1970s. Test equipment gained access to the loop through special trunks attached to a switch in order to perform a series of electrical measurements to see if a problem existed.[23] When loop electronics were installed in the loop, a new channel test was designed to check the electronic components. The channel test was combined with the standard copper test to provide a full loop test. The copper test checked the connection from the end of the loop electronics to the customer premises and became known as the *drop test*. In midband networks, channel tests and drop tests are still required. With broadband, the telephone company establishes a demarcation point between its network and the customer's internal wiring at the NIU. The channel is no longer dedicated to a specific customer's loop because it is assigned at call set-up and it is used to test the facility as a whole and the connections at the NIU.

Test activation can also change with broadband networks. The switch activates the test in narrow-band networks. While this has the advantage of being able to temporarily suspend call processing and use the switch's innate ability to translate the telephone number to the proper loop, it uses a costly interface circuit and takes up to 2 minutes to complete a test. OSSs are used to activate the tests. With midband and broadband, testing can be done while calls are in progress.

There is often an advantage to using the customer's test patterns to locate trouble. In a detailed study of error performance, AT&T found that lines selected for further study on the basis of high in-service error rates were frequently error-free when taken out of service and stimulated with a random test pattern. When the lines were put back in service, the errors returned. A piece of user data can be captured and inserted into a test driver to re-create the problem. This test driver is then run thousands of times to isolate the problem. The pattern can then be incorporated into the regression test suite so that it does not reappear in subsequent releases of the network element software.[24]

Midband ISDN loops have been difficult to install, a fact that suggests the continuing difficulty of maintaining them. Testing ISDN service is more complex than testing traditional analog telephone service because of the many ISDN options and loop constraints. ISDN requires different wiring than POTS and often subtle compatibility problems must be located when terminal equipment and switch suppliers are different. To get ISDN service to work, equipment configuration must be checked, wiring must be correct, and the network must be able to support the service. This balance is so difficult to achieve that Hewlett-Packard created a test approach which is used also for installation.[25]

Protocol analyzers tuned to ISDN are effective troubleshooting tools. Bit error rate monitoring is often used when ISDN circuits are out of service. Known test patterns are sent through a circuit and returned to the test equipment through special loop-back circuits or captured at the receiving end with another test set and differences between the transmitted and the received patterns are noted. In-service surveillance monitoring of ISDN circuits use both the protocol analyzers and the bit error rate measurements to verify that the service is working properly.[26]

With the widespread use of fiber-optic cables in the broadband networks comes the real possibility of noninterference testing. Test systems can send test signals on a fiber cable at a different frequency than the one containing the customer data. Copper networks cannot do this because electrons interfere with each other whereas photons do not. Fiber monitoring, testing, and rearrangement is a viable concept. Using WDM and a small optical switch, it is possible to remotely test individual fibers.

4.3.9. Relative Difficulty of Diagnostic Methods

Figure 4.6 shows the relative difficulty of the diagnostic methods and the location of alarms, camp-ons, and test insertions on a loop. The degree of difficulty corresponds roughly to the relative costs of these methods; as the difficulty increases, there is a commensurate increase in costs. The easiest method is surveillance monitoring where equipment is installed to monitor traffic and congestion and simply detect alarms; the customer is undisturbed. The network intelligence of broadband makes this very affordable. The customer is also undisturbed when the surveillance measurement method is used. Equipment camp-ons at a selected point grab customer data, which are then analyzed using protocol analysis hardware.

The third method in increasing difficulty and cost is test monitoring. Here, a test is injected at the CO and the network is observed for any alarms by using monitoring equipment that is already in place on a broadband network, though often not on a copper network. The customer must be out of service for the duration of the exercise. The most difficult method is test measurement, an alternative that is

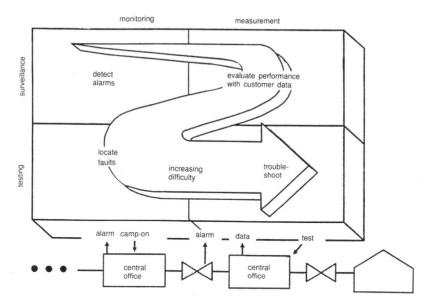

Figure 4.6. Relative difficulty of network diagnostic methods.

used only when all else fails to isolate a trouble. The customer is out of service for an extended period and a test is done that requires elaborate choreography to coordinated technicians at both test point and camp-on point. A test is injected at the test point and data are gathered at the camp-on point for analysis. The test does not traverse the entire network, so if the results are inconclusive, the test must be repeated section by section through the network.[27] The repair center saying goes, "It takes 2 days and 20 minutes to resolve a problem for an uninstrumented network—2 days to find it and 20 minutes to fix it."

4.3.10. Broadband Demands Different Management

A broadband network can be managed by traffic flow, not alarm flow. At line rates of DS-3 and above, measurements can take one of two forms: a directed line-monitoring scheme using the error detection features of the DS-3 framing standards or, if an intelligent network element such as a digital cross-connect is in the path, a simple readout of the performance registers. Technicians who do test measurements need this information, and these techniques eliminate the long and expensive circuit rearrangements required for direct testing.

Perhaps the most promising alternative to direct testing is taking measurements with surveillance access. Digital circuits with directed monitoring devices could capture the customer's data and examine the latter with protocol analyzers. This approach is the way LAN managers tend to approach their jobs. Bachman[28] identified the top five requirements for protocol analyzers: real-time monitoring, protocol decoding, packet capture, packet filtering, and traffic generation.

These same features are being extended to broadband network managers. The best analyzers strike a balance between ease of use and data capture. A major problem with software-based protocols is that they often lag the traffic they are measuring. As traffic grows, hardware-based analyzers that guarantee packet and error capture need to be integrated into the network manager's toolkit.

REFERENCES

1. Katz, M. *Technology forecast*, Version 7 (Price Waterhouse, Menlo Park, CA, 1997), pp. 437–462.
   Engel, J. "Delivery of broadband multimedia services using digital subscriber line technology," speech at LOOP Roundtable (International Engineering Consortium, Chicago, 1997).
2. Ash, G., and Chang, F. "Management control and design of integrated networks with real-time dynamic routing," *Journal of Network and System Management* **1**:3, 237–253 (1993).
3. Bernstein, L. "State-of-the-art broadband access network operations: Beyond flow-through operation," *1996 annual review of communications*, Vol. 49 (International Engineering Consortium, Chicago, 1996), pp. 435–440.
4. Gillespie, A. *Access networks: Technology and V5 interfacing* (Artech House, Boston, 1997).
5. Schwartz, R., and Kreutzer, S. "Broadband networks demand operations support," *Communication Engineering and Design* June 1995, pp. 92–102.
6. Bell Laboratories. *Engineering and operations in the Bell System* (Bell Telephone Laboratories, 1977), pp. 518–523.
7. Kenny, J. "NYNEX reinvents the provisioning process," *Telephony* April 10, 1995, pp. 20–30.
   Aherns, M. "Soft Dial Tone service activation," NOMS '92 (*IEEE* Piscataway, NJ, 1992), pp. 0336–0345.
8. Yavelberg, I. S. "Human performance engineering considerations for very large computer-based systems: The end user," *The Bell System Technical Journal* **61**:5, 765–796 (May/June 1982).
9. Boël, B., McDonald, J., and VanNyen, W. "High speed digital services call for new surveillance techniques," *Trends in Telecommunications* **10**:1, 36–37 (1994).
10. Goldstein, A. J. "A directed hypergraph database: A model for the local loop telephone plan," *The Bell System Technical Journal* **61**:9, 2529–2554 (Nov. 1982).
11. Berkowitz, G. "Operations efficiencies of broadband access networks," *The International Symposium on Subscriber Loops and Services Proceedings* (The Institute of Radio and Electronics Engineers, Australia, 1996), pp. 210–215.
12. Kreutzer, S., and Schwarz, R. "Dynamic times call for dynamic operations," *Telephony* March 6, 1995.
13. Goldstein.
14. Mansouri-Samani, M., and Sloman, M. "Monitoring distributed systems," *IEEE Network* **7**:6, 20–30 (Nov. 1993).
15. Bernstein, L., and Yuhas, C. M. "Net diagnostics: Terms of confusion," *Network World* **4**:27, 27–28 (July 6, 1987).
16. Manber, U. "Chain reaction in networks," *IEEE Computer* **28**:10, 57–63 (Oct. 1990).
17. Bouloutas, A. T., Calo, S. B., Finkel, A., and Katzela, I. "Distributed fault identification in telecommunications networks," *Journal of Network and Systems Management* **3**:3, 295–312 (Sept. 1995).
18. Katker, S., and Geihs, K. "A generic model for fault isolation in integrated management systems," *Journal of Network and Systems Management* **5**:2, 109–130 (1997).
   Bell Laboratories. *Engineering and operations in the Bell System* (Bell Telephone Laboratories, 1977), pp. 563–564.
19. Bell Laboratories, p. 564.
20. Bernstein, L., and Kaplan, F. "Monitoring digital loop carrier," *Proceedings of the IEEE Infocom '87* (The Computer Society of the IEEE, March 31–April 2, 1987), pp. 251–255.
21. Berkowitz, G. M., Fuher, P. T., Gray Jr., B. E., Johnston, A. R., and McElvany, G. L. "A nodal operations manager for SONET OAM&P," *Integrated network management II* (Krishnan, I., and Zimmer, W., eds.) (Elsevier, Amsterdam, 1991), pp. 403–411.

22. Inumaru, Fumio, Sato, Nasohi, Wakabayashi, and Murakami. "Error performance estimation for digital circuits using in-service monitoring information," *Proceedings IEEE Global Telecommunications Conference*, Nov. 30, 1989, Section 51.2.1.
23. Dale, O. B., Robinson, T. W., and Theriot, E. J. "Mechanized loop testing design," *The Bell System Technical Journal* **61:**6(Part 2), 1235–1255 (July–Aug. 1982).
24. Bernstein and Yuhas.
25. Hewlett-Packard. *ISDN Testing Techniques Application Note 397-1* (Rockville, MD, 1990).
26. Smalley, M. "ISDN testing in the real world," *Telecommunications Americas Edition* **31:**8, 36–37 (Aug. 1997).
27. Simonson, T. "All systems GO," *Telephony* Nov. 3, 1997, pp. 54–56.
28. Bachman, D. "Making the diagnosis with Windows protocol analyzers," *Network Computing* **8:**21, 121–132 (Nov. 15, 1997).

5

Pillar—Network Condition

Broadband technology changes the architecture of the telephone network and adds a new dimension to network management. Traffic observation is more vital to broadband network management than alarm monitoring, and service performance, not just network performance, shapes the customer's perception of excellence. There are special problems of transformation between protocols and of scaling networks to greater traffic loads. Techniques invented for ATM are useful in broadband management, as is object-oriented technology.

Wireless communications are ultimately dependent on the network that supports the base stations, and, however popular and growing in customers wireless may be, the high-speed cable network is what will handle data reliably.

5.1. INNOVATIONS THAT SHAPED NETWORKS

There were problems to solve in order to offer new services. How could bursty computer data and voice communication be carried in one network? How should permanent and switched virtual circuits and virtual path connections be managed? Should there be variable length packets or fixed length cells? How should recovery of lost information be handled? The innovations that follow answer those questions and raise more about what a network should be.

5.1.1. Asynchronous Time Division Multiplexing

Narrow-band networks are really a combination of narrow-band POTS service in the loop and broadband switching and transmission in trunks. Dr. Alexander Fraser showed how Asynchronous Time Division Multiplexing (ATDM) could gracefully handle bursty computer traffic and voice communication in one network.[1] His work demonstrated the feasibility of broadband networks for voice, video, and data. He also studied how these networks could be extended and used within the home. This work laid the foundation for the move to Asynchronous Transfer Mode (ATM).

5.1.2. Asynchronous Transfer Mode

ATM is a cell-based way of carrying voice, data, and video traffic on a single broadband network. This is simply stated, but Mischa Schwartz's extensive work in broadband integrated networks explains in detail how ATM works.[2] Managing the permanent and switched virtual circuit and virtual path connections in ATM networks with QoS assurance provides the core bandwidth services that underlie all other services.[3]

5.1.3. Ethernet

Ethernet can route variable length packets, and Fast Ethernet switches are used for some broadband communication. However, these applications will not scale to larger, faster networks because Ethernet was designed for use only as a LAN technology. For this reason, Ethernet provides neither positive acknowledgments nor priority.

ATM technology is the basis for faster networks than Ethernet because it is a cell relay technology. Cells are fixed-length packets. Special-purpose hardware can be designed to process cells quickly and it is easier to monitor cells rather than variable-length packets as they progress through the network. To gain speed in networking, software-based packet switches evolved to become routers by separating the processing of control information from data.

5.1.4. Internet Protocol

There is a fundamental incompatibility between ATM and TCP/IP in how they treat the loss of an ATM cell. The entire TCP/IP packet of 400 to 1500 bytes may be retransmitted, rather than just the dropped cell. Therefore, the realizable throughput of the network has been shown in simulation to be as low as 37%. For this reason, the engineered design load of packet networks should probably be no more than 40%.

Routers based on the Internet Protocol (IP) do not set up connections, they cannot ensure that packets will arrive in sequence, and they drop packets in heavy traffic. They merely look at data packets and use a routing table to determine which router gets the next packet. This packet-by-packet switching was included in IP to make the network available even in the face of multiple failures without an external congestion management OSS. Although this is a desirable attribute, the QoS is more difficult to ensure in IP networks than in ATM networks because of the lack of discrimination in the order in which packets are routed. IP democratically provides only one level of service quality for all users. Some enhancements are being designed that will permit an application to pick from several available QoS levels.[4]

Both of these constraints present a vital network management issue: Congestion must be assiduously avoided. Congestion triggers the loss of cells or packets and the problem is then amplified by the retransmission of full packets. There is ongoing study of how to map TCP/IP over an ATM network.

5.1.5. Best Bets

As Fred Brooks is fond of saying, there is no silver bullet. Each solution has benefits and drawbacks. There is an emerging effort to revisit WDM as the core technology. Cells and packets could be mapped atop wavelength slots. The control problems associated with packet switching that interfere with the achievement of acceptable QoS for nonbursty voice and video have some researchers looking at new uses for WDM that reserve some slots for ATM, some for IP, and some for the circuit switch. The thinking is to avoid the problems of a fully integrated, full-service network by not trying to maximize the merging of bursty with a stream of traffic. Instead, the traffic is separated. This is an engineering trade-off.

For these reasons, broadband networks must be based either on ATM technology or, once the differences are resolved in how ATM and IP handle addressing, on a combination of WDM and ATM. Some researchers are exploring new switching technologies while others have added an IP stack to ATM.[5] No matter which backbone protocols emerge to become standard, it appears that ATM will probably have shaped the broadband network management techniques. The choice of technology depends on the profile of offered load for a specific application and the ease with which it can be managed.

5.2. BROADBAND MANAGEMENT CONCEPTS

In order to manage a broadband network, one must have some way to keep track of widely distributed network elements. One must also develop a suitable metric, specific to the characteristics of this domain, for problem identification. Finally, one must be able to link physical and logical relationships together. The discussion in this section addresses these needs.

5.2.1. The Role of Object-Oriented Technology

Management of the ATM networks embraces the use of object-oriented technology. As network elements are, by their nature, widely distributed, distributed object-oriented technology using classes of managed objects is ideally suited to the problem. The managed object is characterized by a set of attributes that reflects the state of the corresponding real object or network element. These objects track configuration information and cell performance internal to the network.

Network management systems for broadband cell-switched networks must control each virtual circuit as allocated resources are modified. They need to monitor call set-up time and control the operating parameters of the signaling system. These and other management functions are encoded in managed objects installed in every network element. Events happen very quickly in broadband networks, so managed objects must capture and process the behaviors of each network element. To rely on communicating the information back to an OSS for processing would saturate the network with management information. Therefore, the attribute values of the managed objects are stored in the Management Information Base (MIB) where an OSS can poll them when needed.[6]

The broadband OSS maintains a high-speed, multiplexed communication interface to each cell switch. It collects performance, accounting, and status information for further processing. It provides standard interfaces so that it can integrate the management of any LANs attached to the broadband network. James Martin describes the architecture and implementations for LANs.[7] LANs support peer-to-peer communication where all communicating elements exercise similar control. This contrasts with hierarchical networks where one element controls the network operation. LANs are the backbone of client/server networks. When LANs are connected to broadband networks, a network management system must be able to control the LANs as well as the broadband network elements.

5.2.2. What Measurement Best Shows Network Problems?

One of the great advances of Datakit,[8] Fraser's cell switch, was the use of software to detect lost cells and errors in cells within the cell switch itself. These ideas have been extended to ATM networks so that they now have self-provisioning, self-test, and automatic error detection features built in. The network switch checks cell headers for transmission errors and discards those cells that might be misrouted. With only modest additional cost, checks on cell payload and discarded cells related to corrupted data were built into Fraser's Datakit. Network monitors collected these data and triggered repairs before customers became aware of problems.

Early in the development of network management for fast networks, it was assumed that bit error rates would indicate problems related to equipment or fiber link failures. Bit error rates were, after all, the workhorse of facility management. However, the quality of the links was so good that networks either worked well or failed completely. Bit error rates that did occur were the result of dropped packets where collisions had occurred. Applications were therefore measuring lost packets as bit error rates. This confusion masked the fact that the loss was related to congestion, not equipment problems. ATM cell loss rates came out of the analysis of this problem in order to reflect what was really happening in the networks. While both are important, traffic observation is more vital to broadband network management than alarm monitoring.

Experience with Starkeeper,[9] which was used to manage cell networks connected to LANs, showed the importance of real-time observation of network traffic. The traffic profiles proved so useful that they became the most important way of observing network behavior. This surprised designers who expected a reliance on alarm processing to find network element failure. That network managers first checked displays showing traffic flow before looking at alarm charts was an important insight pointing to a fundamental change in the way broadband network management systems need to be designed.

There is huge investment in alarm processing, so traffic monitoring is often the poor cousin, which makes network management systems more expensive and less efficient than they might be. The shift to managing traffic as the first line of defense and using alarm management as reference information is an idea driven by broad-band networks, but it is applicable to all aspects of network management. At the AT&T National Network Control Center and in regional control centers, experienced network managers observe traffic and respond to congestion conditions first and then look at alarm information to find the source of a network element failure.

5.2.3. Plant Engineering Database

When carriers engineer their networks for 95% fill to minimize capital investment, they find that they have a serious ongoing expense problem. There is not enough slack for efficient maintenance operations and for easy provision of new services. To keep costs down, a company must balance the debt ratio with the number of employees per 1000 lines.

Service providers can change their POTS provisioning approach to accommodate new services by merging their physical outside plant data records with logical data relationships. This data structure is used in conjunction with new engineering methods to compute the electrical characteristics of particular loops. The Local Exchange Carrier (LEC) may be able to choose loops that satisfy the customer's specifications. A model of the physical plant includes the following:

- Customer identification.

- Connectivity of the equipment into a logical circuit.

- Electrical properties of the circuit.

- Status of the circuit in terms of its present use and future use.

- Points where density of plant use is measured, called *taper codes*. They measure the fill rate or use rate of the pairs of wires within a cable.

- The point of actual circuit connection of customer equipment in relation to all possible connection points. Typically a pair of wires may be connected to a customer at any of several locations.

- The location of the taps along the circuit.

- The build of the circuit from *sections of plant* connected at splice points to local assemblies of the cables and pairs connected in *laterals* at cross-boxes or terminals.

- The characteristic of the terminals and any special provisions for wiring out of POTS limits.

The physical model together with its logical extensions are packaged in a relational database and an engineering application is applied to the data structures to determine suitability of the plant for the customer circuit. Special interface software is used to extract information from other logical databases when this database is not used for all engineering and assignment purposes. The application has the ability to recognize, track, and mark physical and logical plant items as they are installed, activated, deactivated, taken out of service, and physically removed from the outside plant. The architecture provides for the use of a single common database across all of the functions or for the use of multiple databases with each dedicated to a particular function.

If special equipment is needed at the customer's location for service operation or testing, a new "left-in" administration process with supporting software is invoked. This "left-in" approach is similar to the dedicated plant concept designed for POTS. Special features are needed for spectrum management to reduce the risk of cross talk in the copper pairs. Though nothing exists at this moment, it would be desirable to have a new OSS to watch for changes in the physical equipment as it monitors construction jobs. It would be able to do this because the plant rearrangements are always reflected in the physical network database. The logical pointers in the physical/logical databases can associate outside plant equipment with a particular circuit, its electrical characteristics, and its customer. If the plant equipment changes, the electrical characteristics would be computed again to ensure that the circuit specifications continue to be satisfied.

5.3. BROADBAND READY

During the transition from narrow-band to broadband, millions of dollars must still be spent on narrow-band plant and legacy OSSs. This money and effort can be directed toward making the narrow-band world "broadband ready." The concept of broadband ready includes the following practices:

1.   Purchase of any new narrow-band plant facilities with the intention of being able to reuse them in the future broadband network

2.   Actively capping legacy OSSs for the diminishing copper base

3.   Focusing reengineering efforts to make narrow-band operations upwardly compatible with future broadband

The third point is particularly important. An orderly evolution from narrow-band to broadband plant is possible only with appropriate network planning. Figure 5.1 shows the progression from the existing copper network where a circuit is established at service installation, through a transition stage of being broadband ready where there is fiber from the CO to a hub, to the ultimate change in the physical placement of equipment to constitute a broadband network.

When narrow-band networks need to be upgraded to meet the immediate needs of customers, the following engineering rules that anticipate the eventual change to broadband are appropriate.

5.3.1.   Replacement

Copper cables must be replaced by composite copper/fiber cables. A comparison of these cables is shown in Fig. 5.2. The outside plant construction work is the

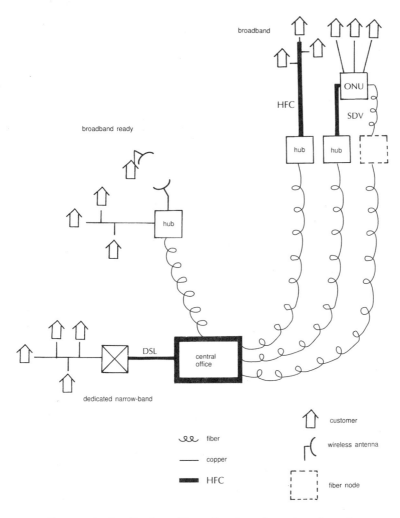

Figure 5.1. Application of broadband ready engineering rules.

most expensive part of upgrading the plant. Placing fiber beside the copper cable prepares the plant for eventual broadband use and takes advantage of the cost of any construction. Figure 5.3 shows the physical labor involved for aerial cable; buried cable presents an even greater problem when street digging permits must be secured and resurfacing is necessary.

5.3.2. Upgrade of Distribution Cables

When feeder or heavy usage distribution cables are upgraded, most service providers use fiber cables. In 1998, fiber is a better economic choice than copper when it serves more than 400–500 customers. Fiber and digital loop carriers are well

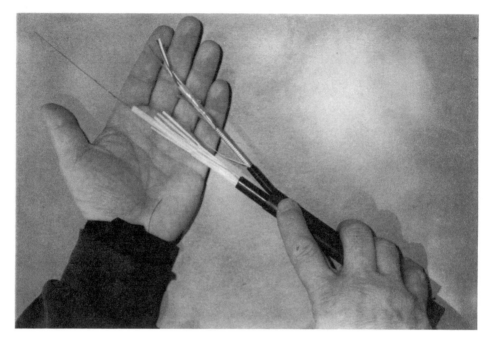

Figure 5.2. Comparison of copper and fiber cable.

Figure 5.3. Making the outside plant broadband ready.

suited to service all the way to high points of concentration such as university and corporate campuses, apartment complexes, and office buildings. For less dense routes, digital loop carrier technology is often used to multiplex many customers' loops on a few copper pairs. As the loop approaches the final destination, copper distribution and dedicated drop wires are used to reach homes or offices. To be broadband ready, the electronic hubs should be sited at points that are best for eventual fiber distribution and not at those best suited for the digital loop carrier. This prepares the site and, most importantly, ensures that there are adequate power arrangements.

5.3.3. Management

In terms of management, the network must be treated as if it were broadband when looking toward the CO and treated as if it were narrow-band when looking toward the premises served by copper. As the hubs move closer to the customer, the narrow-band systems are gradually replaced.

5.3.4. Midband Service

For midband service, ADSL equipment may be placed at the hubs, the CO, or even at a special server. The choice is an economic one driven by the need for and location of POTS splitters. Outside plant connections to the CO must use the V5 or TR303 (more recently known as GR303) standards for TSI.[10]

Broadband-ready networks are well suited for managing unbundled services and telephone number portability. Figure 5.4 shows the transition in more detail (note: TR08 was an early device for providing interface into a switch, but it did not provide for dynamic TSIs). The constraints imposed on services by the narrow-band loops and the cost complexity of managing these services make broadband networks attractive. When service providers can reduce the cost of provisioning and maintenance, broadband becomes even more compelling. The prime obstacle to moving forward with broadband is the practical difficulty of obtaining building permits to dig up communities and the cost of restoring private property after construction.

5.4. EVOLUTION TO BROADBAND

A protocol called *routing to intelligence* allows for regionally based services, rather than forcing each switch to have a common set of features. Figure 5.5 shows how adding SONET rings to the access network allows wireless, broadband, unbundled loops, and TR303 TSIs to get past the MDF. An intelligent router in front of the voice switch recognizes from the packet headers that they are going to various servers and bypasses the voice circuit, obviating the need to convert to voice circuits and reconvert to packets to go out again. The intelligent router sends the packets directly to the server via the SONET access ring. The overall concept of routing to intelligence is beginning to appear in various partial implementations.

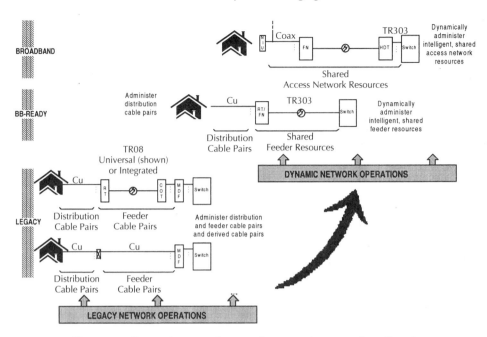

Figure 5.4. Dynamic network operations span legacy to broadband.

For example, some ADSL manufacturers are not integrating with the circuit switch and can split out voice, using a dedicated line going to a router.

Network management features demonstrate the feasibility of such a system with the complex collection of OSSs that must work together to operate a network. Broadband makes this even more attractive as it reduces the latency in such networks. The management of a network with distributed functionality resembles the type of system administration needed to manage large mainframe computers. ATM can be used to build this network by locating a fast ATM switch at each CO. Voice calls would then be sent to the traditional switch while data and video sessions would be routed to their appropriate servers. Each local ATM switch would have a name manager for connecting the customer with the appropriate server. The speed of maintaining these databases becomes vital for successful broadband implementation. Emergency (E911) and other lifeline services are peeled off the local lines and immediately sent to their servers so that this vital network function continues to operate.

The trunks become permanent virtual circuits in an ATM network. This high-powered "turbo-trunk" concept allows the ATM routing algorithm to reroute traffic automatically in the event of congestion or network element failure. SONET overhead channels can be used to carry signals or commands for setting up virtual circuits and eventually become permanent virtual circuits themselves. Today's thinking is that ATM will be modulated within SONET channels. However, the advent of WDM may obviate the need for SONET altogether. In that case, ATM would run as the ATDM signal on fiber channels. Consistency in integrating various technolo-

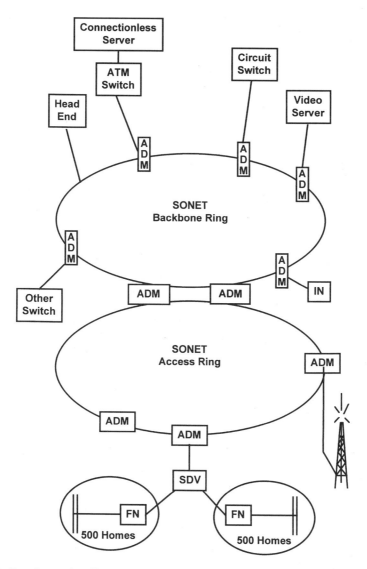

Figure 5.5. Routing to intelligence. ADM, Add/Drop Multiplexer; IN, Intelligent Network made up of intelligent routers; FN, Fiber Node; SDV, Switched Digital Video. Attached to the circuit switch is either an MDF or a direct interface.

gies to carefully layer management functions could make all of this possible, though no disciplined effort has yet been made in this direction.

5.5. CONVERGENCE

By combining dynamic provisioning with routing to intelligence, service providers can use their narrow-band switches as intelligent network elements, leaving it to the new ATM line terminations to convert customer messages into digits and packets. These switches would continue to support voice services, including such functions as providing dial tone and busy signals, but other services would bypass them to go directly to the server of choice. By placing a specialized ATM switch that is instrumented with network management features in every wire center, narrow-band and broadband services can gradually merge. The telephone numbering plan and information services can be maintained through this transition. This is not a conservative, incremental solution, but it is highly logical for long-term benefit. An alternative solution to the number portability problem would use centrally controlled databases in the call path (like 800 service control point architectures) to let customers keep their phone numbers as they change service providers. Although this approach appears to be simpler, it actually seriously impacts the OSS infrastructure and the geographic nature of the COs. It would lead to slow call set-up and horrendous consistency problems in the huge databases that would be required. Another alternative is to continue to upgrade the voice switches with features. The problem with this is the amount of time it takes to develop the software and put it into the switch; in addition, exceedingly complex software would be needed.

5.6. INTERNET TELEPHONY

Most of the popular PC packet phone software leverages the fixed monthly charge for Internet use. Demand for low-cost long-distance calling has driven advances in speech compression technology to improve the quality of packet phone calls. In addition to the voice transmit and receive operations, packet telephony systems must also set up and end calls, query directories, identify themselves, and report problems.

QoS is an important limitation of packet telephony. The perceived voice quality depends on the compression coding techniques and the bit rate. Broadband networks often experience noise bursts which are particularly distressing because voice compression packs more speech information into fewer bits than an uncompressed signal. Delays caused by switch latency and jitter, variations in delay, affect QoS.[11] A packet telephony transmitter tends to send a packet at a roughly constant rate while one customer is talking. The receiver expects to process these packets at that same rate. When packets do not arrive on time, the customer listening will hear noise, garbled speech, or, worst of all, dead air. Delays can occur in packet networks when bursts of data traffic occupy most of the packets, because packet networks gain their efficiency by not reserving capacity for each active call. When packet telephony is used to send fax, these problems are minimized. The challenge is to meld *computer communications-based* packet network management techniques with *circuit-oriented* public telephone network management techniques. Techniques for reserving bandwidth for voice calls need to be developed because the routes through the network vary from call to call and congestion is seen at router buffers handling all types of traffic. This becomes even more important for video calls. Congestion on narrow-band loops and QoS concerns for Internet voice may stimulate the move to broadband networks.

5.7. FEDERATED SERVICES

The explosive growth of the Internet stimulated the development of new services such as interactive shopping, home banking, and electronic commerce. These services are "federated," depending on an infrastructure that spans multiple independent networks. Managing federated services is difficult because only a small part of the customer network can be observed and controlled by any one service provider. Service contracts among providers are often fashioned to make management possible. Diagnosing problems in federated networks is hard because of the diverse mix of network elements, the lack of experience on anyone's part, and the wide-ranging, unpredictable nature of failure modes. The faults observed in the initial deployment of one federated network spanned network element failures, application incompatibilities, user errors, software bugs, and incorrect client configurations.[12]

There were similar problems in another federated network.[13] Sometimes customers downloaded useless files, senselessly truncated because their browsers put an arbitrary limit on the size of file transfers. Another problem was poor implementation of Internet protocols, resulting in random garbled messages. Prodigy truncates e-mail subject heads to extremely short lengths, which were not anticipated by some companies selling information by e-mail. Software bugs and configuration problems are hard to trace, but human factors problems are truly a challenge. Almost half of all customers make mistakes when they type their own e-mail addresses into a merchant's Web site. Broadband network managers need tools to deal with these service-driven problems in addition to tools that handle conventional network problems.

The practical implication of such interactions is that the symptoms of a problem can appear far away from the problem itself, both in space and in time, and may have no apparent correlation with the problem. The customer, however, sees it as a failure of services. Diagnosing these problems requires information about all of the components on which a service relies, but this may not be available if the components are in a domain that does not export information outside the domain. Yemini has discussed similar phenomena in the context of large-scale heterogeneous networks, focusing on the problem of alarm correlation.[14]

Good support determines the quality of the customer's experience with complex Internet services, particularly for unsophisticated users, yet that care becomes especially difficult in a federated network. First Virtual Holdings, an Internet-based electronic commerce service company, found that the biggest unexpected problems centered on customer service, and that "an Internet-savvy customer service department is an absolute prerequisite for anyone providing commercial services to the net."[15]

Economics for the mass market dictate that customer support departments cannot rely solely on human expertise to handle the growing complexity as well as increasing numbers of users. Service managers need ways to ensure that their Web sites are responsive. Just having bigger pipes to the site is not enough; the traffic flow needs to be controlled. A client downloading a huge file can block a customer wanting to make a purchase. A tool set that can assign bandwidth to different kinds of transactions will emerge eventually, containing such tools as a load balancer that redirects transactions from a busy or broken Web site to an available one. Balancers will eventually load servers as a function of the amount of computer resources available versus the congestion in the network going to the server.[16] Once loads are under control and customers get a high level of service, the service manager must provide extensive directory services. These would include domain name services that link the Internet service provider to the customer.

A network management system for these federated systems would have to tackle the tough issue of multiple suppliers. The design would necessarily focus on diagnosing individual services and not on diagnosing the entire network. This is a major change in viewpoint for service providers who try to manage only within their boundaries, but it is a necessary change because customers care about end-to-end service. A management system would have to separate diagnosis and customer care from service itself and coordinate contracts to specify the expectations of a service. Contracts would also define the responsibilities of each service provider and create an infrastructure for enforcing compliance. It is important that a neutral third party be identified who can arbitrate in case of conflicts. The tools used to install and verify the service need to be available to diagnose problems that may occur.

Incrementally enhancing centralized management systems with Internet technology will not scale to handle the growth of federated systems. Only a distributed architecture can separate management logic from management data and therefore accommodate growth. The network that is being managed carries within itself the network management data. Research at the University of Florida investigated the best way to get access to network data on the Internet.[17] Three models of data distribution were studied: conventional centralized, partially distributed, and fully distributed. In each case, the study calculated the total time required to gather the needed management data and the load that those data placed on the managed network. The results showed that a partially distributed data model reduced the load on the network demanded by a centralized architecture. It reduced intermediate storage requirements and eliminated the unacceptable delays experienced with a fully distributed model. The importance of having the network become the network management database is highlighted in this study and provides a road map for broadband network data management architecture.

5.8. ACCESS METHODS

Midband services, two-way video and telephony services, and high-bandwidth symmetric services require different access methods. The following sections, 5.8.1 through 5.8.3, treat these methods.

5.8.1. DSL Options

Service providers may choose to offer midband services with DSL systems that exploit the transmission capacity of twisted copper pairs, as was explained in Section 2.2, as a means of making a reasonable transition to high-speed networks. These technologies allow service providers to offer 1.544 Mbps or faster service over existing copper pairs.[18] DSL in its simplest form consists of two transceivers, one at each end of a loop. With the addition of a POTS splitter at each end of the loop, any failure in ADSL equipment is prevented from affecting voice service, therefore keeping vital lifeline and emergency telephone service working even in the face of prolonged power outages. Splitters offload DSL frequencies onto a separate copper pair at the customer site; keeping data apart from voice frequencies reduces line noise during high-speed data transmissions. Although configuring DSL without a POTS splitter is cheaper in terms of equipment and personnel, it does sacrifice speed. George Hawley, an early pioneer in digital loop carrier, wrote definitively describing the history and architecture of the varieties of DSL, known collectively as xDSL.[19] By 1998, the suppliers of DSL equipment provided many options, each having some unique advantage. The options are arranged in Table 5.1 in ascending order of data speed to the customer.

The problem is that each of these solutions requires severe engineering tradeoffs to fit within the constraints of the copper loop. These networks are managed using the traditional narrow-band approaches with the services mapped into virtual circuits. This mapping complexity makes provisioning very expensive and complicates surveillance and testing. The customers' best choice depends on their location and their planned use of the midband data communications. For applications with extensive file sharing, customers need the HDSL or SDSL versions.[20] With all of the choices and the critical dependence on the electrical properties of the available loops, expensive special service provisioning must be used because POTS provisioning is not appropriate.

5.8.2. Hybrid Fiber Coax

HFC combines coaxial and fiber-optic cable. Cable TV companies have been replacing coax with fiber to eliminate many of the amplifiers required to keep the signal at an acceptable level. These amplifiers are usually designed only for transmission to the customer so they limit the return path and therefore the effectiveness of Internet sessions. When fiber is extended to the neighborhood nodes, a single coaxial cable serves several hundred homes. This architecture eliminates many of the maintenance problems for cable TV service and expands bandwidth. With HFC in place, a cable TV company can easily provide telephony service.

Telephone companies are also interested in deploying HFC technology. Southern New England Telephone is spending $4.5 billion to recable Connecticut for two-way video and telephony services using HFC. Michael Heller outlines multimedia service and trials for 37 service providers worldwide.[21] Despite all of this activity, copper will remain in the local loop for a long time. Broadband investments offer both the telephone and cable TV companies the opportunity either to reduce the cost of providing service or to make their QoS acceptable. Detailed business case analysis shows the need to consider life cycle and not just first costs when deploying a broadband network.[22] HFC can economically provide either video or telephony services alone, but its strength comes from providing a full-service two-way network.[23]

Table 5.1. Digital Subscriber Loop Options

Option	Speed to customer	Speed from customer	Comments
Integrated Services Digital Network DSL (ISDN/DSL)	128 kbps	128 kbps	Uses mature ISDN technology
Symmetric DSL (SDSL)	768 kbps	768 kbps	Supports equal transmission in both directions for IP telephone calls
Splitterless DSL	1 Mbps	90 kbps	No need for costly field installation of a splitter at customer location. Carries risk of cross talk
Consumer DSL (CDSL)	1 Mbps	128 kbps	Easier to deploy than ADSL because line cards go in existing switches. Carries significant performance penalty
High-bit-rate DSL (HDSL)	1.544 Mbps	1.544 Mbps	Works well at short distance from CO. Very sensitive to condition of copper plant
Rate-adaptive DSL (RADSL)	Up to 6.3 Mbps	Up to 640 kbps	Service provider must activate the desired bandwidth up to the capability of the loop[a]
Asymmetric DSL (ADSL)	9 Mbps	640 kbps	Maximizes amount of data that can be sent from Internet to PC, consistent with POTS
Very-high-bit-rate DSL (VDSL)	53 Mbps	2.3 Mbps	More of HDSL

[a] Carter, W. "Keeping pace in the high-speed race: WebSprint rolls out rate-adaptive DSL solutions," *Telephony* September 1, 1997, p. 22.

HFC architectures use fiber to carry video and telephony from the head-end at the cable TV video server or CO to an optical node serving a geographic area. When cable companies add telephony to their cable networks, they also add self-identification and self-test equipment. These additions make it possible for cable TV network managers to rapidly pinpoint network problems. Without this equipment, cable TV companies wait for multiple customer reports to pinpoint problems. This leads to long lags between problem reporting and repair and over time created the public perception that cable TV is unreliable.

However, cable TV is a shared medium. When there is low individual traffic, there is no problem. The specter of unreliability might arise again when traffic grows and the QoS degrades as many customers try to share the same medium.

The cost of upgrading the HFC is shared by telephony and data services. Improved network management makes it attractive to prove-in telephony on HFC. As the cost of photon-to-electron conversion decreases, the fiber nodes will move closer to the customers and eventually to the curb. In addition, the use of passive optical amplifiers will increase the distances that fiber-optic cables can reach without a corresponding decrease in reliability. The goal is to attain telephony levels of reliability without incurring a $15–$20 per line cost. Embedded network management and no active network elements from the head-end to the customer make this goal achievable.

Once HFC broadband is in place, providing two-way telecommunications requires adding the proper plug-in. Plug-in return capability already exists in the optic nodes. Self-identifying plug-ins can be used so that expensive inventory management systems will not be required. Some cable operators are already using return transmissions for surveillance.[24] Cable TV companies need spectrum management for the return path and HFC network element management systems for the installed HFC in order to get dynamic provisioning and self-maintenance. Telephone companies are including these functions in their first deployments of HFC.[25]

5.8.3. Switched Digital Video

SDV extends fiber to the curb. For telephone companies, fiber optic cables come from the CO to a node in the neighborhood serving about 30 homes. This is a more expensive solution than HFC for providing broadband in the access network but handles video calls extremely well.

SDV networks carry ATM cell-switched digital, video, data, and Internet traffic. Analog cable TV is transported over a one-way fiber-coax overlay network that also carries the power for the neighborhood nodes. Managing cross talk, customer traffic, bit rates, spectrum, and noise bursts are important for good QoS and will be part of the network management system.[26] SDV is a good step toward moving fiber as close to the customer as service providers can afford. The durability and low maintenance of fiber in the outside plant makes this attractive. Passive optical networks are the ultimate in low-maintenance networks because they have no active elements from the CO to the customer.

SDV is best suited for high-bandwidth symmetric services. The ONU installed in a neighborhood can house built-in network management features. SDV offers the telephone companies the opportunity to reduce their provisioning costs through the use of dynamic provisioning.

5.9. SLEEPING GIANT

Utility companies are well positioned to provide broadband service. They have rights of way, extensive customer data, technicians, billing systems, and a substantial communications network in place. Some utilities have been installing fiber cables along with new electrical cables reaching into neighborhoods. The challenge facing utility companies is to gain experience and expertise in operating a high-QoS telephony network. Broadband network management offers them the opportunity to have many management functions embedded with the network elements and to rely on customer care and trouble reporting OSSs to handle customer service. Such expansion of their business base allows utilities to earn more revenue and keep customers loyal for their electric service as deregulation of the electric industry takes place.

Utility companies are installing two-way customer communication systems to read meters, detect outages, and reduce peak energy consumption by managing the customers' appliances. Broadband networks are an attractive way of implementing two-way customer communications. These networks can also be used for management of the electric lines, cable TV, and telephony. Some utilities may look to become the provider of modern electric and telephony service with broadband network management techniques applied to both.

REFERENCES

1. Fraser, A. G. "Early experiments with asynchronous time division networks," *IEEE Network* Jan. 1993, pp. 12–26.
2. Schwartz, M. *Broadband integrated networks* (Prentice–Hall, Englewood Cliffs, NJ, 1996).
3. Anerousis, N., and Lazar, A. "An architecture for managing virtual circuit and virtual path services in ATM networks," *Journal of Network and Systems Management* **4**:4, 425–455 (Dec. 1996).
4. Blumenthal, M. S. "Unpredictable certainty: The INTERNET and the information infrastructure," *Computer* **30**:1, 54 (Jan. 1997).
5. Vaughan-Nichols, S. J. "Switching to a faster INTERNET," *Computer* **30**:1, 31–32 (Jan. 1997).
6. Yemini, Y. "A critical survey of network management protocol standards," *Telecommunications network management into the 21st century* (Adiarous, S., and Plevyak, T., eds.) (IEEE Press, New York, 1994), pp. 19–24.
7. Martin, J. *Local area networks* (Prentice–Hall, Englewood Cliffs, NJ, 1989).
8. Fraser.
9. Becker, D., Liss, W., Weinstein, A., and Wilson, E. "Starkeeper® II NMS-management of a cell relay network," *AT&T Technical Journal* **73**:4, 46–55 (July/Aug. 1994).
10. Gillespie, A. *Access networks: Technology and V5 interfacing* (Artech House, Boston, 1997).
11. Houghton, T. F., Schloeme, E. C., Szurkowski, W., and Weber, W. P. "A packet telephony gateway for public network operators," *ISS'97: Proceedings of World Telecommunications Congress* (Toronto, Canada, Sept. 1997).
12. Bhoj, P., Caswell, D., Chutani, S., Gopal, G., and Kosarchyn, M. "Management of new federated services," *Integrated network management V* (Lazar, A., Saracco, R., and Stadler, R., eds.) (Chapman & Hall, London, 1997), pp. 337–340.
13. Borenstein, N. S., *et al*. "Perils and pitfalls of practical cyber commerce," *Communications of the ACM* **39**:6, 36–44 (1996).
14. Yemini.
15. Borenstein.
16. Bernstein, L., and Yuhas, C. M. "And the walls come tumblin' down," *IEEE Communications Magazine* Dec. 1992, pp. 126–133.
17. Anderson, J., Ilyas, M., and Hsu, S. "Distributed network management in an INTERNET environment," *IEEE Global Telecommunications Conference* (IEEE, Piscataway, NJ, 1997), pp. 180–184.
18. Price Waterhouse. *Technology forecast: 1997*, Version 7 (Price Waterhouse World Technology Centre, Menlo Park, CA), pp. 110–114.

19. Hawley, G. "Systems consideration for the use of xDSL technology for data access," *IEEE Communications Magazine* **35:**3, 56–60 (March 1997).
20. Lawson, S., and Kujubu, L. "Nortel, Rockwell add CDSL voice support," *InfoWorld* **19:**47, 45 (Nov. 24, 1997).
21. Heller, M. "Technology and business choices for residential multimedia services," *Multimedia over the broadband network: Business opportunities and technology* (International Engineering Consortium, Chicago, 1996), pp. 40–43.
22. Arellano, M. "The economics of infrastructure upgrades," *Multimedia over the broadband network: Business opportunities and technology* (International Engineering Consortium, Chicago, 1996), pp. 19–23.
23. Mulla, H. D. "HFC network applications and design issues," *1997 annual review of communications* (International Engineering Consortium, Chicago, 1997), Vol. 50, pp. 527–531.
24. Paff, A. "Hybrid fiber/coax in the public telecommunications infrastructure," *IEEE Communications Magazine* **33:**4, 40–45 (April 1995).
25. Heller.
26. Aprille, T. J., Schwerin, L. M., Sipes, J. D., and Stevens, N. S. "Interactive broadband services and PCS network architecture," *Bell Labs Technical Journal* **1:**1, 11–27 (Summer 1996).

6

Pillar—System Environment

It was becoming increasingly clear by 1990 that independent development was creating islands of isolated OSSs that were unable to work together in any integrated way. The very same cottage industry mentality that plagued early software programming had once again created muddled, unwieldy confusion, this time of ominous proportions. The push to decompose OSSs into layers was an effort to impose some conformity to standards. The Telecommunications Management Network (TMN) standard was developed by professional organizations with the goal of allowing equipment from many suppliers to operate under a common network management umbrella. It separated OSS functions into element, network, service, and business management layers.[1] Older legacy OSSs were split into element management and network management subsystems, taking some guidance from the TMN standard. TMN is considered to be a network separate from any telecommunications network and needs to be managed itself, which adds another dimension of complexity.[2]

Unfortunately, there has never been any agreement about what constitutes the corpus of functions of the element management layer. Because that basic agreement on the nature of any system's foundation does not exist, element management systems are often bundled arbitrarily with network elements. Sometimes element management systems handle a collection of similar network elements, and other times they take on some network management control functions. This is because suppliers of network elements often consider element management systems as added value to their products and give them to customers free of charge as inducements to buy. These element management systems then act as adjuncts to the network elements. Such arbitrary packaging makes conformance to the standards difficult. Administration of these complex, feature-rich element management systems requires substantial software assistance. This trend will continue and accelerate as vendors pragmatically build interfaces to legacy OSSs and call them element management systems, no matter what they actually do.

Systems can now include one network element, the product line of one vendor, or a common type of network element produced by several vendors. As there is no agreement on the set of intelligent functions, some may take on, for example, the functions of thresholding and control, while others do not. Competitive cost pressures force each vendor to optimize its collection of network elements and interface abilities to a competitive price point. This leaves a flawed foundation for subsequent layers and affords little integration of information at the level of linking OSSs.

As the network management technology matures, standard interfaces between the service management and the network management layers must evolve. Such interfaces will be services-based. The service management layer will not need knowledge about the underlying network technology to provision a service; it will command the network management layer to configure network elements consistent with the needs of the service using a predefined set of parameters. The network management layer will set up the connections and automatically activate the service. Detailed differences between network elements and the element and network management layers will absorb the network configuration options.

6.1. MARKET PRESSURES CAUSE RETHINKING OF OSSs

Competitive pressures have forced service providers to integrate computer and telephony technology to provide better customer care and to increase the effectiveness of marketing. This forced the reengineering of business processes and led to major investments in the service management and business management layers.

Figure 6.1 shows the migration of telecommunications functions and the likely trend in the next decade toward business management systems. As a result of the migration of functions, element management systems manage equipment, network management systems manage the paths across the network, and service management systems manage what goes over the paths. Business management systems execute company policies and the rules with which services are implemented. Figure 6.1 shows levels of abstraction that isolate various components with object oriented technology. Without the technology to allow the decoupling of complexity implied in these layers of abstraction, networks are limited to only the degree of complexity that the human mind can hold. The sophistication of decoupling complexity is necessary to achieve the management needed for broadband networks. If the information-hiding philosophy of that technology is violated, small changes will then ripple through many layers with unknown effects.

Most large carriers now have many complex OSSs that together rival the complexity of the network being managed. Existing OSSs do not work well together. There are often major functional overlaps, different user interfaces, and different ways of representing the same data. This state of affairs is the result of the evolution of individual OSSs from different viewpoints. Attempts to reconcile this hodgepodge into a "system of systems" failed because the underlying software technology was not up to the task of partitioning functions, and this resulted in cascading problems.

What follows are descriptions of three different approaches to managing network elements: network element centered, work center focused, and functional module creation.

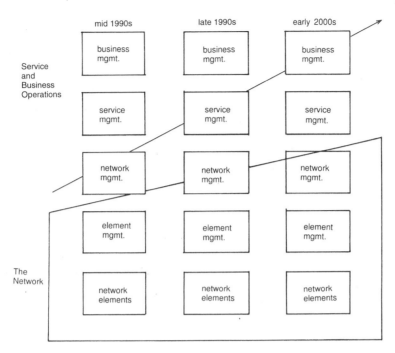

Figure 6.1. Migration of telecommunications functions. Customer demands change the focus from network management to service and business management.

6.1.1. Network Element Centered

In the network element approach, the design focuses on a particular network element around which is created a specialized multifunction system. Such systems are highly desirable when there are only a few types of network elements to manage. This approach consolidates craft expertise, is the least expensive way to support the network element, and lets the OSS developer focus on the built-in features of the network element. It is a low first-cost solution. The problems come when there is a need to centralize the work force or handle many different types of network elements. Then technicians are faced with a desk full of terminals and computer centers must run many isolated OSSs.

6.1.2. Work Center Focused

In the work center approach, work responsibilities spanning an operational discipline or a geographic area are defined. A work center is a group of people assigned to one aspect of providing service; examples are telephone assignment and repair centers. Several very successful OSSs have been implemented from the perspective of optimizing all of the functions of a work center. The Loop Management

Operations System (LMOS)[3] is an example of how well this approach can work. The OSSs first provided machine aids to the people doing the work; then full flow-through automation was achieved. Eventually, these task-focused OSSs collided with those that were network element centered as telephone companies tried to integrate their OSSs.

6.1.3. Functional Module Creation

The goal of the functional module approach is to centralize all similar soft-ware functions into one system, such as all surveillance tasks in one system and all provisioning tasks in another. The functional module approach can cross geo-graphical areas, as in the case of GTE doing alarm monitoring for all of its systems from Dallas, Texas. The objective is to decrease the number of overlapping systems and make it possible to use one craft terminal for all functions. This reduces system administration costs and eliminates desktop clutter.

Network management suppliers provide the ability to give widespread access to functions. The functional module approach often leads to easy planning but demands a high degree of standardization on functional partitioning. However, stan-dards are sometimes abandoned when the pressing need to get on with business is combined with delays in the development of the software constituting such com-prehensive systems. Work center managers have little patience for the promises of the total system in the face of meeting customer commitments. Delays force service provider managers to turn to interim systems based on the network element or work center approach to accomplish specific tasks. Once installed, these interim systems tend to stay on long after the interim problem has passed; they become entrenched in the business operation.

There is only one major example of the functional module approach working well, and it is not a model for future development. Bellcore's Operations Systems Strategic Plan (OSSP) integrated a variety of administrative systems for telephone company assignment and provisioning for POTS, special services, and, ultimately, some broadband. Bellcore was in a unique position to be successful because first, they benefited from the legacy of the overarching control and standardization that had been imposed by the old Bell System, and second, they were therefore able to enforce standard interface specifications called *contracts* between element man-agement systems and between OSSs. When Bellcore attempted to build its own element management system, it encountered the same problems that plagued the rest of the industry; the standard interfaces avoided the whole issue. These contracts were later used industrywide.

OSSP was successful because it was able to define both the conceptual archi-tecture and the environment and control the functional allocation within the plan. OSSP is not a likely model for the future because it had its origins in the Bell System, a controlling management force that was able to interpret standards and keep them viable, and such a force no longer exists. When similar schemes have been attempted subsequently, as in the case of TINA-C, which will be discussed in Section 6.7, they have depended on the cooperation and goodwill of all suppliers. Rapidly changing technologies and rapidly changing regulatory environments mitigate against the ongoing success of such enterprises.

6.2. RECONCILING THE THREE APPROACHES

Always there is embedded in the architecture of OSS software a model, implicit or explicit, of how tasks will be performed. The approach used in the design of the OSS (one of the three just described) strongly influences the model. OSSs based on different models are difficult to reconcile to each other. Functions are inevitably duplicated, data structures are often different, and user interfaces proliferate. Resolving these differences is not important when the business situation permits islands of mechanization to exist with no impact on efficiency. However, when tasks are reengineered into new cross-disciplinary processes to improve customer service and reduce operations costs, OSSs must be interfaced together to provide fully automated flow-through. Then the feathers hit the fan.

6.3. ADMINISTRATIVE DATA IN THE NETWORK

Networks work better when network elements contain administrative information that describes how they will be used. Including such descriptive data resolves the common discrepancies between the record inventory and the physical inventory of network equipment. Much time and money are lost in upgrading a network when engineering plans cannot be followed because it is difficult to keep track of what is installed.

Those who manage public and private networks are faced with the daily challenges of maintaining operations within parameters set by their employers as well as by customers, suppliers, and government entities. The need to keep the business running well has made these managers advocates of standards. They cannot afford to become confined to a single supplier and therefore constrained in upgrading network management tools as new equipment and services become available.

Open Systems was the rallying cry of the beleaguered in direct response to IBM's proprietary SNA, which was centered on the mainframe. It was difficult for system administrators to incorporate the emerging departmental minicomputers and PCs into their networks. Special tools had to be obtained or developed, and most administrators reverted to simple problem detection schemes and protocol analyzers monitoring specific lines to fix problems. This was detective work, not systematic engineering. It limited the growth of corporate networks, added costs, and, most critically, undermined the accuracy of corporate databases because users could not update them as quickly as they could solve business problems.

6.3.1. Open Systems Interconnection

Network managers turned to standards organizations to define approaches that vendors were expected to follow so that there would be some hope of manag-

Table 6.1. Layers of OSI Reference Model

Layer number	Layer title	Functions
7	Application	Access to OSI environment and distributed information services
6	Presentation	Resolve data format differences
5	Session	Logical connections and session connections
4	Transport	Data flow and transmission errors
3	Network	Packet-forming and routing
2	Data link	Sharing transport and low-level error handling
1	Physical	Interface to physical medium

ing heterogeneous networks. An internationally standardized solution to this problem is based on the seven-layer Open Systems Interconnection (OSI) Reference Model. This model classifies communications tasks into seven layers as shown in Table 6.1. Each layer performs a different function and this schema became the foundation technology for the platform approach to building network management systems.

This model lets the transmitting computer send data down its stack from the application layer to the physical layer, while the computer that receives the transmitted data reconstructs the latter from the physical layer up the stack to the application layer. This is done independent of the application and separates details of the intermachine communication from the function desired in the application. For network management applications, this is a key step in being able to add new devices and management techniques while the network continues to operate. Most client/server applications use the Internet Protocol (IP) at layer 3 and the Transmission Control Protocol (TCP) at layer 4, transport. Applications using TCP/IP for interfaces conform to layers 3 and 4. The power of this approach is that different transport media can be used simultaneously in the network without the individual network applications being aware of it.

6.3.2. Router Capacity

With the rapid growth of Internet traffic came the need for routing capacity of many gigabits per second. Average IP packet size doubled to 1500 bytes in 5 years as a result of use of the Web. As an ATM cell holds 53 bytes, 5 bytes of which is header, the ATM overhead gets added more than 30 times to each Web inquiry. This has led to studies of IP switching and gigabit routers.[4] IP switches can be combined with ATM by using a pool of permanent virtual circuits containing the overheads involved in making connections. Routers can be scaled up from LAN packet switches using connectionless network protocols suitable for high-speed IP routing. Network elements of either technology will appear in broadband networks and need to be managed.

6.3.3. Functional Areas of Network Management

Standards groups recognized five functional areas for network management: configuration, fault, performance, accounting, and security. Configuration management involves the inventory and connectivity of network elements. Fault management is concerned with reducing network downtime. Performance management monitors filtered network data that cross thresholds to indicate the health of the network. Accounting management provides for organizational chargebacks and capturing network costs. Finally, security management is vital to protecting the network owner's investment in hardware and software, as well as the integrity of the network and application databases, against penetration, piracy, or unauthorized alteration.

6.3.4. Protocols

Protocols convey the network management data and commands. Telephony people spend great effort standardizing on OSI agents and their managed objects. Their protocol is called the Common Management Information Protocol (CMIP). When communication is required between several open systems, it is handled by the Open Systems Management Process component of CMIP, residing at OSI level 7. Each network element defined to a network management system, called a *managed entity*, is accessed through CMIP and can have an MIB attached to it.[5]

Client/server people, on the other hand, use the Simple Network Management Protocol (SNMP). SNMP uses MIBs that are different in structure and content than the CMIP MIBs. SNMP is a data-transport protocol that is part of the suite of TCP/IP protocols.[6] It is the standard management protocol for TCP/IP networks created by the Internet Engineering Task Force. Implementation of SNMP places some development effort on vendors of network elements to include managed objects in their products (e.g., multiplexors, LANs, routers). One tool in the SNMP tool kit permits the creation of communications capabilities for the managed element, called *agents*.

OSSs can use a collection of objects built on platforms as a foundation for building TMN-compliant systems. The TMN concept of a manager and agents for the management functions reflects the physical separation of *managing* operations systems from the *managed* network elements. Manager applications typically add value to network information in a specific domain through analysis and inference and have an integrated view of the network of managed objects. Agent applications, on the other hand, have intimate knowledge of the elements of the network that they are representing. These network elements may be basic components such as switches and multiplexors, or they may be other management systems that are acting in manager roles themselves for other domains or other views of the network objects.

Table 6.2 lists the messages SNMPv1 employs. The SNMP protocol uses a client/server query–response model. The client is called a manager and generates the queries. The server is called the agent and generates the responses. Managed

Table 6.2. SNMPv1 Messages

Message	Function
Get Request	Asks for one or more object variables
Get Next Request	Asks for the object variable following this one
Set Request	Updates one or more object variables
SNMP Trap	Agent alerts server that an event has occurred
Response	Agent responds with the data requested
SNMPv2 adds the following:	
Get Bulk Request	Fetches a large amount of data (multiple MIB variable values)
Inform Request	For manager-to-manager communication

nodes contain the agents and have the capability to go beyond the client/server model, giving the agent the ability to notify the manager when certain events happen. The management servers execute processes that gather data and issue commands to the agents. Each agent contains a set of access routines and the managed object for the network consistent with the MIB format. The servers use the Abstract Syntax Notation One (ASN.1) language to allow them to manage network elements without regard to the computers used. ASN.1 is important because it defines data types and object values independent of any particular bit encoding technique. This makes the MIBs vendor neutral.

Proxy agents with SNMP interfaces can be used for network elements that do not have built-in agents when time is short or development resources scarce.[7] Agents or proxy agents talk to the network management system using SNMP. If a platform is used, the SNMP software handler in the platform distributes the messages to the appropriate application. Translations from proprietary network management messages common across many elements to SNMP are done by the element's MIB. Version 1 of SNMP (SNMPv1) is widely deployed, while version 2 (SNMPv2) is in its early stages of use. Unfortunately, the SNMP upgrade is not totally backward compatible. SNMPv2 adds security features, but they are very complicated, and bulk retrieval of agent data can cause congestion. Other features of SNMPv2 are as follows:

- Origin identification

- Message integrity

- Limited replay

- Access control

- Data confidentiality

- Bulk retrieval

- Manager-to-manager interaction

6.3.5. Conformation to OSI

Both CMIP and SNMP conform to the spirit of OSI layering concepts, but they do it differently in the details. OSI is conceptually fine, but it is too complicated. Even TCP/IP protocols do not use all of the OSI layers because of the overheads implied by strict conformity. There is a fundamental conflict between CMIP, which tries to conform to OSI layers from the perspective of a telephony background, and TCP/IP and SNMP, which are more pragmatic and try for efficient implementation.

CMIP supports the notion of out-band data; SNMP supports in-band data. CMIP accepts unsolicited data from the network elements; SNMP polls for data. CMIP carries a higher overhead burden than SNMP. CMIP scales to handle large networks but is too expensive for small networks. SNMP is efficient for small networks but does not scale easily.

Most network management systems need to manage on the order of 10,000 nodes. The issue is to be able to grow beyond that without having to replace the network management systems. It is possible to have systems that manage 80,000 or more nodes. Beyond 80,000, network designers will have to partition the networks.

Nevertheless, SNMP support is essential in telephony network management platforms so that networks can be managed from end to end. Market trends show that customer premises devices, LAN/WANs, computing infrastructure, and even PBXs are being managed using SNMP. Telecommunications service providers and customers also have large complexes of customer premises devices that must be managed, and they will use SNMP for this purpose.

With two types of network management servers and two types of networks, four interfaces for complete connectivity are needed. The complexity grows when signaling and message networks of several varieties are added. Unfortunately, it is not possible to choose one way. The OSI base is too expensive for simple networks and SNMP is inadequate for global telephone networks.

SNMP is best used to manage customer premises data networks and to send data about service provider networks to customers for incorporation into their network management systems. It is best suited for monitoring and controlling routers and bridges that connect LANs to LANs, and for high level exchanges of management data such as trouble administration and configuration data. Structured queries to large management information databases are best done with OSI protocols.

6.4. TMN ARCHITECTURE

TMN is designed to support analog and digital networks. It is equally applicable to service provider networks or companies having their own private networks.

It can be used to manage transmission or switching systems, entire exchanges, circuit and packet switched networks, signaling systems and terminals, customer terminal equipment, and telecommunications support equipment.

6.4.1. Logical, Physical, and Informational Parts

The TMN architecture has three parts. Its *logical* part specifies the management functions and the reference points for data exchange between the functions.[8] These logical functions were discussed in Section 2.11. The *physical* part defines how these functions are implemented on real systems and the interfaces between them. The *information* part defines the data structures. This part is based on the ISO Systems Management Model and defined using the Guidelines for the Definition of Managed Objects (GDMO). In addition, the Transaction Language 1 (TL1) protocol defined by Bellcore in 1984, is the most widely used protocol for tying OSSs together.[9]

TL1 is based on two simple concepts. First, the interface should be readable by humans as well as by computers, and second, the identification of the data is carried along with the value of the data as information is transferred between systems. This very powerful approach allows systems not needing all of the data elements in one set of information blocks to still use the interfaces and uncouple the systems where designers want them uncoupled. TL1 protocols can be mapped into object mediators and have been easier to build than CMIP ones. Above all, the point is never to build protocol software—buy it instead.

6.4.2. Monitoring the Network

TMN can monitor network elements in nearly real time. Using the information it obtains, TMN determines the nature and severity of the problem. It can control aspects of the network elements on demand while monitoring their performance. It can help locate failures, assist in installation, and help bring equipment into service. The following are the five TMN management function areas:

- Fault management
- Configuration management
- Accounting management
- Performance management
- Security management.[10]

TMN usually uses a separate network from the one being managed to communicate the management data. The use of the term *network* in TMN signals that the management is performed by a cooperating collection of OSSs rather than by a single monolithic one.

6.5. APPLICATION OF TMN

The telecommunications network can be thought of as a distributed application. Consider a model describing the telephone network as a worldwide application that delivers end-to-end service under user control. Too often, the management of this network is based on a static model with fixed location of functions, a high degree of central intelligence, and a single protocol. Designing and building TMN interfaces is one of the most difficult software jobs. It requires detailed protocol experts who also understand the scope of the complex of systems used to manage the networks. To complicate things further, uncertainties in the OSI standards cascade into delays in the TMN work. The Network Management Forum has defined a set of detailed specifications called Open Management Interoperability Points (OMNIpoint) to be used to solve real business problems in a way that matches TMN standards.[11]

TMN may be implemented in a four-layer client/server architecture as shown in Fig. 6.2.[12] The figure shows different OSSs used for each layer. If one OSS spans several layers of the TMN interface, then the X interfaces are not needed because data are passed among code within applications. The drawback of a single OSS is that the layers are no longer discrete, and the clean separation that TMN layering tries to achieve is destroyed.

Figure 6.2 shows standard interface points with specified interface protocols. One of these is the important Q3 interface used for data exchange between OSSs and network elements. Q-adapters are used to interface with noncompliant systems.[13] These are difficult to build because of the complexity of mapping between the TMN-defined interfaces and the noncompliant ones. Legacy databases may be accessed with a combination of standard query interfaces, e.g., Standard Query Language (SQL), file transfer-type interfaces, e.g., Bellcore's contracts, and terminal emulation. Interfaces between OSSs use the TMN Q3 interface that follows the OSI stack and provides connections between applications.[14] The layering partitions the OSS functions into four layers from top to bottom with standard interfaces between the layers. The element management layer may or may not be architecturally distinct from the network elements. Using high-performance computer processors, network element vendors bundle the element management functions directly in the network element, especially for simple network elements. The database or MIB of the element manager is intimately shared with the network element using automatic data replication technology. Often the interface between the network element and the element manager is a proprietary high-performance one, tuned with the Q3 interface to the network element with a standard interface between the network management layer and the element manager.

6.6. A SUGGESTED CLIENT/SERVER ARCHITECTURE FOR LINKING OSSs

The OSSs rely on a common user access layer to provide technician access to data and controls. The OSSs are designed with an Application Programming

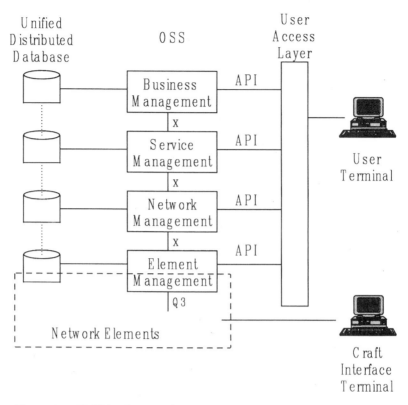

Figure 6.2. TMN implementation of four-layer client/server architecture.

Interface (API) that connects them to the user access layer. In the client/server architecture, the user access layer relies on a LAN collection of servers providing the desktop options of dumb terminals, workstations, and PCs for the technicians.

The data for the OSSs may be stored in a common database rather than having them distributed functionally with each OSS. The inability to share data in a common way adds costs creating situations where large amounts of network information must be entered manually or through expensive special translation software and coordinated among various systems. Data often become inconsistent and unusable. Distributed database technology lets designers keep common data models between applications without suffering transaction delays to get the data when they are needed. The data describing the network configuration and the network elements' capabilities can be based on an object-oriented model that provides a single data structure.

The network management systems, service management systems, and business management systems are designed around core disciplines. They perform no *data steward* functions but have restricted read/write permissions to the common database.

Software platforms are commonly used for the operation and administration of the client/server OSS in a distributed computing environment. The platform provides the tools and libraries for building standards-based communications interfaces. The use of software platforms promotes large-scale software reuse and modularity, which decreases system first-costs and future maintenance costs. The software platforms make it possible to protect investments in hardware platforms through support of hardware independence and distributed processing. Availability of cost-effective software platforms is a key to meeting the management and support needs of network managers.

The platforms provide the tools for creating managers and agents and maximize automation of those parts of the development process that are driven by an object definition phase. The MIB contains object class definitions, object instances, and attribute data for the objects contained in the models that define the network. These data are stored so as to be readily accessible to manager and agent functions. Sometimes data are replicated to ensure access, but in that event replication database technology ensures data consistency.[15] The TMN standards are written using object-oriented specifications and because the object-oriented technology promises more cost-effective and faster solutions,[16] the next-generation OSSs model the network as a collection of objects. It is not easy for application developers to become proficient in object-oriented technologies, however, and creating a team of such programmers represents major time and educational investment.

6.7. EVOLUTION, NOT REVOLUTION

The key requirements for telecommunications interfaces and their associated object models are *interoperability*, *integration*, and *flexibility*. The existing TMN defines an information architecture that is focused primarily on interoperability. An extensive tool kit supports an architecture made up of the CMIP, the GDMO, and the OSI systems management functions.

Integration and flexibility are another matter. The Open Distributed Processing Reference Model is the basis for the TINA-C architecture. It can be used to migrate TMN to a dynamic architecture operating in a distributed computing environment. It yields a rigorous specification of interoperable interfaces, eliminates the protocol barrier preventing integration of different systems, and gives flexibility in service provisioning.[17] An important objective of the TINA-C effort is to control the cost of operating multivendor networks, because the cost of managing a service tends to increase exponentially with the size of the network.

When applying the Open Distributed Processing Reference Model to TMN, the selection of tools is critical. That selection is one of the products of applying the TINA-C guidelines. A transport network is decomposed into a number of independent networks, each with characteristic information concerning a client/server association between adjacent layer networks. In turn, each component network can be partitioned so that TINA-C guidelines can be applied to manage the network at the level of detail desired.[18] TINA-C borrows layering concepts from TMN, but uses them differently. It combines the network and element management layers into a

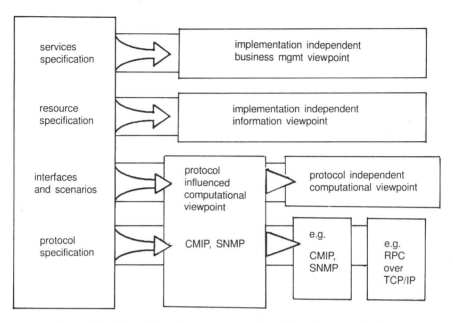

Figure 6.3. Translation from conceptual model to implementation.

single layer. TINA-C complements the TMN approach by taking a service-oriented, top-down view to managing network components, whereas TMN has a bottom-up resource focus. Both are needed to effectively manage broadband networks. Figure 6.3 shows practical ways of realizing network management. Conceptual models are translated into various types of information that are necessary to run a networking business.[19]

Without the addition of TINA-C, CMIP models are difficult to implement because the resource specification, interface specification, and implementation of scenarios are bound into the protocol specification. This makes it difficult to reuse the specifications for implementation using other protocols. Because protocols use various data structures and have differing capabilities, the current computational specifications are inevitably influenced by protocols selected for engineering viewpoints. For example, the CMIP protocol incorporates a database query for filtering the results of inquiries. A series of gateways is possible for run-time interaction to translate from a generic API to one based on CMIP. These gateways can make TMN converge with a distributed computing approach, if it provides protocol transparency, integrates existing management protocols through APIs, and provides a framework for component reuse.

Successful TMN-based object models provide the interoperability that makes rapid service provisioning possible. Distributed system techniques used in concert with TMN in TINA-C deliver the integration and flexibility benefits.

6.8. FIELD TRIALS IN GERMANY

Multiple broadband multimedia field trials have been conducted in Germany for more than 10 years.[20] One objective was to understand how these services should be managed. Another objective was to test and evaluate how best to assemble services for multimedia products according to different customer profiles. The residential customer market and the business customer market were studied separately. In the trials, the following applications were built on a multimedia services platform:

- Teleworking

- Telemedicine, integrating patient folders in multimedia documents

- Telepublishing, combining the work of various editorial offices and print shops to shorten production time and facilitate distribution

- Tele-education

- Telecomputing, collaboration of distributed work forces for maintenance

- Tele-information, systems handling bookings and providing tourist information

- Video-on-demand services

- Telecommerce, banking, and shopping

6.8.1. Trial Parameters

Different networks were used for the field trials. Most business multimedia applications rely on symmetric communication patterns. Although individual sessions may be asymmetrical, their direction may be either way depending on caller and purpose. Business multimedia projects used an ATM network platform. The pilot network was extended beyond the national borders by links into the European ATM pilot, a network supported by 17 network operators in 15 European countries. Management of ATM connections was conducted from a central site. Network management software was developed to support configuration and fault management of virtual path connections. The software communicated with ATM switches with a Q3 interface. It was found that multiple network management systems needed to be linked to manage the ATM permanent and switched virtual circuits crossing service provider boundaries. The TMN standard needed to be extended to allow this linking to occur.

For residential customers, different applications were tested. Practical considerations dictated using existing physical access networks. This constrained the available bandwidth for a large portion of private homes and small offices. Consequently, test scenarios were designed to minimize additional infrastructure investment. One was a hybrid approach, where cable TV networks provided a downstream channel to distribute video data, and the upstream channel used the existing telephone network. Another was a pure cable TV-based solution using HFC technology. Fiber-optic cable is used in the most concentrated sections of the CATV distribution plant with coaxial copper cable attached to reach the home. Repeaters in the cable TV network were upgraded to handle bidirectional traffic. A third approach was to use the existing twisted pair copper telephone access network upgraded with ADSL technology.

Different forms of interactive television were tested. Digital video was stored in central servers. Individual downstream video channels were established on customer demand within the network, and an individual return channel from the user to the server infrastructure was used to customer control the delivered service. At the CO, the video was merged with the voice telephony with a POTS splitter. On the customer's premises, another POTS splitter diverted the telephone channels from the video stream and command channels, the latter to be decoded in an ADSL modem that linked to the set-top box. There, audio and video streams were distributed to the terminals, and control sequences were fed into the return channel. Where ADSL is employed, an MPEG coded video stream of 2048 Mbps and a 9.6-kbps command channel can be transmitted over distances up to 3.5 km.

A management system controlled the smooth operation of the server and distribution processes. It remotely administered the elements of the network. Another system administered the customer base, content available in the servers, and the usage parameters necessary for billing the requested services.

6.8.2.   Lessons Learned—Business Environment

Making complex software work together and creating the protocol stacks needed for business applications was much harder than anticipated. Trial-and-error troubleshooting demanded patience from all participants. Furthermore, all of the network elements needed to be of very high quality.

Problems were observed when several components needed to work together to support real-time human collaboration applications. These used video, audio, and data at the same time. These were handled in different ATM channels. Tests traced the troubles to processor overloads occurring in user terminal equipment when handling video, audio, and data streams simultaneously. Aligning the audio with the video is very challenging to the network management systems. Managing LANs at both ends of the ATM connections was very difficult. Two different implementations of IP were tested. A propagation time difference of two to three was noted. This makes network planning sensitive to the particular protocols used. Administering client addresses was a difficult task.

Bursty IP traffic had to be handled over ATM constant-bit-rate virtual circuits. Investigations showed that the ATM in-house systems employed by pilot customers just mapped IP packets into ATM cells and sent them sequentially. The ensuing bursts would need more bandwidth than allocated for the respective constant-bit-rate IP service initially offered. Protocol measuring test devices were needed to catch this.

Another traffic-sensitive problem saw cells surpassing the contracted peak cell rate being discarded by the public network portion. Whenever a cell carrying part of an IP frame is discarded, the TCP protocol retransmits the whole IP frame. The result was to reduce the effective transmission speed. The consequential introduction of traffic shapers to restrict cell rates to the contracted peak cell rate solved this problem.

Great efforts were put into a European field trial for cooperative TMN interface development. More time than estimated had to be spent on interoperability tests of TMN protocol implementations, an experience also confirmed by earlier similar work.

Management applications must evolve to accommodate heterogeneous management protocols and network structures in backbone and access networks. Traffic management methods need to deal with B-ISDN/ATM networks including PVC and SVC traffic of variable bit rates. Multimedia traffic patterns are still unknown to a large extent, but they will heavily influence traffic management.

6.8.3. Lessons Learned—Residential Environment

The ADSL technology was stable and no interference with telephony services was detected. The modems employed offer the capability for remote maintenance and diagnosis. The equipment components were set-top boxes, ATM switches and application, and video server worked reliably, but users found the configuration unwieldy and intrusive. Intermittent software interface failures had to be debugged when the trial was started.

Effective management for residential multimedia networks requires more knowledge about typical usage patterns. Management of the newly introduced video and information server structures requires further studies, especially how they scale as the services grow. The traditional assumption of a typical 3-minute telephone call used for designing voice telephone networks does not hold for multimedia networks.

6.8.4. Clear Need to Manage

Though the objectives of the study were to introduce a variety of possibilities into business and residential arenas and see what had utility and appeal, the results were inconclusive except to demonstrate the technical possibility of the enterprise. The study did show a clear need to put in place techniques to manage the installation and ongoing operation of the broadband network. Configuration problems were particularly challenging. Neither this study nor any others have been able to identify what the driving market forces might be for broadband, calibrate the demand for it, or suggest a possible exploitation of its market potential. The difficulty of the technology and the discomfort level of installing and using it are still too high.

REFERENCES

1. Bernstein, L., Davidson, C.W., and Rad, C. "Functions, attributes and fields of application for the telecommunications management network," *Proceedings Globecom 1987*, pp. 1264–1267.

 2. Hall, J ed. "Management of telecommunication systems and services: modellling implementing TMN-based multi-domain management," *Lecture Notes in Computer Science* (1996).
 3. Schenker, L., and Barbera, S.J. "Automated repair service bureau," *The Bell System Technical Journal* **61:**6, (Part 2) (July/Aug. 1982) pp. 1095–1096.
 4. Newman, P., Minshall, G., Lyon, T., and Huston, L. "IP switching and gigabit routers," *IEEE Communications Magazine* **35:**1, 64–69 (Jan. 1997).
 5. Terplan, K. *Communications networks management* (Prentice–Hall, Englewood Cliffs, 1987).
 6. Hegering, H.-G., and Yemini, Y. "An introduction to the simple management protocol," *Integrated network management III* (North-Holland, Amsterdam, 1993), p. 261.
 7. Barbier, S. "Systems management in the 1990s," *AT&T Technical Journal* (July/August 1996). **73:**4, 82–97.
 8. Special Issue: Telecommunications Management Network, *Journal of Network and Systems Management* **3:**1 (March 1995).
 9. Man, F.-T. "History of TL1," *Journal of Network and Systems Management* **7:**2, 143–148 (1999).
10. Shrewsbury, J. K. "An introduction to TMN," *Journal of Network and Systems Management* **3:**1, 13–38 (March 1995).
11. Hall, p. 127.
12. Byrne, C. J., Raman, L. G., and Woo, H. "Realizing a TMN," *Journal of Network and Systems Management* **3:**1, 39–72 (March 1995).
13. Glitho, R., and Hayes, S. "Telecommunication management: Vision vs. reality," *IEEE Communications Magazine* **33:**3, 47–52 (March 1995).
14. Wikkala, T., *et al.* "TMN X interface testing for pan-European network management," *Proceedings, NOMS '96,* pp. 274–277.
15. Desai, S.R., Follett, D. J., Sinha, G., and Sundarmurthy, C. "BaseWorX™ platform for building object-oriented management applications," *Globecom '93 Technical Program Conference Record,* IEEE Catalog No. 93CH3250-8, pp. 194–202.
16. Snell, M. "Client/server development talk turns to object tools," *Software Magazine* Oct. 1994.
17. Lengell, M., Pavon, J., Wakano, M., and Chapman, M. "The TINA network resource information model," *IEEE Communications Magazine* June 1995.
18. Hamada, T., Kamaata, H., and Hogg, S. "An overview of the TINA management architecture," *Journal of Network and Systems Management* **5:**4, 411–435 (Dec. 1997).
19. Special TINA edition, *Journal of Network and Systems Management* **5:**4 (Dec. 1997).
20. Bartsch, F.-R., and Auer, E. "Lessons learned from multimedia field trials in Germany," *IEEE Communications Magazine* **35:**10, 40–45 (Oct. 1997).

7

Pillar—Software Functions and Features

Rejuvenating access plant is a constant process. This investment in access technology is driven by the need to accommodate new services, improve service quality, and reduce operational costs. One's competition always threatens to offer something dazzling or to operate on a lower cost curve. The replacement of access infrastructure provides an ideal window of opportunity for modernizing OSSs and for reengineering the underlying operations processes. The combination of capital investment and standardization of business practices is a necessary feature of software success.

Unlike copper access networks of the past, modern broadband access networks are characterized by intelligent network elements. The feature of built-in intelligence supports dynamic provisioning, performance monitoring, and capabilities for fault isolation. By coordinating the design of intelligent network elements and OSSs, it becomes possible to reengineer the way access networks and services are managed to gain efficiencies and improved service that would have been impossible in isolation.

When networks are made up of complicated systems, the software must function to allow data management across the network. Thinking in terms of database management is no longer effective. Finally, the software must function as specified. This is an area that requires much continuing study because software is now only conditionally stable and has no theoretical basis to explain its behavior. There are conscious management decisions that can be made, however, to mitigate the effects of the chaotic nature of software.

7.1. LIBERATION FROM LIMITATIONS OF THE PAST

Section 4.1 described the legacy of telephony as a manual process. Hard-won though it has been, flow-through is simply automation of that process, not a rethinking of it. The process still requires frequent interruption for inside and outside plant work, and data can be contaminated as errors are made, closeouts are reported incorrectly, or plant is rearranged without updating the operations database. Dedicated inside plant and dedicated outside plant (DIP/DOP) policies were meant to alleviate some of the problems, but limited technology made this approach so unwieldy that these policies were unenforceable. The limitations of flow-through are mostly associated with the lack of intelligent operations capability in the access network elements.

Now technology has liberated the process from mere mechanization to become synergistic interaction by allowing network elements to be self-aware, self-monitoring, and self-healing and by allowing the OSSs to derive information directly from the network elements. These capabilities and reengineered processes together with a new OSS infrastructure allow dramatic improvements in operations with equally dramatic reductions in operations expenses.

Previous switch/access interfaces, such as TR-08, required each working line to have a physical assignment at the DS-0 level on the switch to a Host Digital Terminal (HDT) interface. The HDT would generally require a second physical relationship between the DS-0 specified on the HDT/switch interface and the physical port at the NIU. These two links were generally coordinated in external databases associated with the OSS provisioning systems. Having such a specific relationship limited the flexibility of the HDT and required the entire process to be tracked by the OSS database. With frequent rearrangements and repair activity, the OSS database and reality would often drift apart, resulting in service provisioning errors.

Major change to the design of OSSs that will support broadband networks is required to achieve operational targets. The new OSS uses a platform-based and standards-based approach to software development to establish an integrated operations infrastructure across a broad range of services and network technologies. This approach also allows the flexibility needed to accommodate variations in local operating conditions as well as different methods and practices associated with different business models. The OSS implements the reengineered process and works in concert with the network elements that make up the intelligent network access.

With the added intelligence in the access network, particularly an NIU at or near the home, service provisioning and network reconfigurations can be performed. Likewise, data regarding network configuration are obtained from the network itself rather than companion databases. Data are synchronized to the network. This provides the opportunity for real-time, on-demand provisioning and the opportunity to provide additional information to customer service representatives.

With broadband, the network becomes the alarm system. Amplifiers, power nodes, fiber nodes, and the like have transponders that automatically generate alarms if problems occur. NIUs are polled regularly for problems. This information is sent to the OSS, which can automatically analyze problems, sectionalize a particular network failure, and dispatch a technician when necessary. This allows proactive maintenance, all of which can occur before customers notice a problem.

7.1.1. Functional Architecture

Overall, the OSS manages several kinds of information concerned with providing timely service and with long-term planning to anticipate network growth. In

Chapter 4, the Service Order (SO) was explained as an external request from a customer. Response time to an SO can range from minutes to days. The corresponding request that is generated internally in the interest of the company is called an Engineering Work Order. Response time to a Work Order can range from weeks to months. Traffic data are continually gathered to serve many purposes in three major categories: subsecond, to manage the network configuration, weekly, to manage configuration changes and anticipate traffic patterns, and monthly, to manage the installation of new equipment. OSS architecture accommodates, tracks, and monitors all of these activities. An example of its functional blocks for a network supporting telephony and video services is shown in Fig. 7.1. The functional blocks are described as follows:

- The *customer satisfaction management* module integrates the customer contact function for telephony and video services. It provides interfaces for customer service representatives to gain access to the provisioning and maintenance process management modules, to use data from the network in the determination of service readiness of the network, and to administer video services from both the end-user perspective and the video server perspective.

- The *work management* module provides support for the management of workflows for the maintenance and provisioning processes. It tracks changes needed to provide customers with services, upgrade the network, and fix problems. It has direct interfaces to engineering for plant redesign. Provisioning process management allocates equipment and loops from available inventory following a predetermined set of rules. Maintenance process management controls testing, trouble reports on loops, requesting new loops from provisioning as needed and triggering technician visits to broken loops.

- The *living unit database* module inventories the connection of a living unit to the network for both telephony and video services. The location of available customer equipment, the type of OP, planned changes to equipment, and other attributes are matched against service attributes to determine the viability of delivering services to the living unit over existing connections.

- The *physical network database* module inventories the physical network elements. Static configuration about the physical plant is stored in this database to support alarm analysis, craft dispatch, and maintenance activity. Dynamic network information is accessed from the network itself or from the element managers to ensure that accurate information is utilized.

- The *data collection* module interfaces to the network for billing and traffic data.

- The *traffic* module sends data to engineering.

- The *billing and account management* module sends data to billing and market analysis.

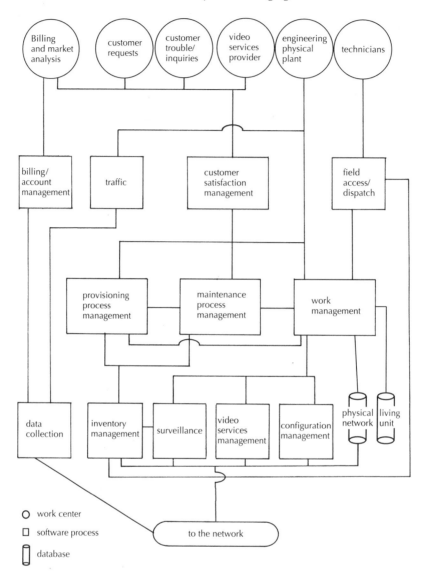

Figure 7.1. Functional blocks of OSS architecture.

- The *inventory management, surveillance, video services management*, and *configuration management* modules, like the data collection module, interface directly to the network. When there are configuration changes in the network, surveillance detects them and notifies inventory management, which updates the physical network database.

- The *inventory management* module tracks critical assets, such as NIUs, through their life cycle. Because this module is integrated with provisioning and maintenance process management and with field access, there is a high degree of integrity maintained in the inventory data. This module includes features supporting customer/supplier interactions such as automatic ordering of warehouse items triggered by stock levels. This information can be electronically bonded to suppliers, facilitating *just-in-time* inventory management procedures.

- The *surveillance* module collects access node alarm data, notes customer reported data, and performs analysis to determine where network failures may exist. This module uses the physical network database module to understand the relationships among network elements and reduces multiple alarms to a single common cause. Work requests are automatically forwarded to the work management module if dispatch is required. To facilitate accurate dispatch, this module utilizes a Geographic Information System (GIS) that can display the exact geography and superimposes the physical plant onto that geographical map. Customer troubles, alarms, and current craft locations are displayed on that map.

- The *video services management* module has two major functions: (1) managing customer service profile across the video elements in the network and (2) managing configuration of the video delivery network.

- The *configuration management* module is responsible for coordinating service activation activity for narrow-band services. It manages the assignment of originating equipment in the switch, time slots in the HDTs, and the relationship to NIUs.

- The *field access and dispatch management* module provides automated assignment and distribution of work requests to technicians in the field based on expert rules and queuing algorithms. The work requests integrate the work description with any inventory tracking information. The information is provided via a field access terminal. From that terminal, technicians can access and close work requests, synchronize inventory with their activities, and configure customer services for both telephony and video.

- In addition to inherent new functionality, interfaces to customers' legacy systems are supported for such upstream activities as billing and market analysis data.

7.1.2. Telephony Provisioning

The legacy telephony provisioning process requires a complex array of OSSs that was originally designed for *multiple copper plant,* cable pairs that can be accessed from one of several places. OSSs have evolved over the years and have been enhanced for Digital Loop Carrier systems. These enhancements were placed on top of the cable pair base and worked reasonably well; however, there were always new services and equipment that required manual work-arounds or manual intervention. The key problem for provisioning flow-through was a complete reliance on database systems that were often out of synchronization with the actual network elements. Every reworking was an occasion for new errors.

Rather than attempt further enhancements, a new OSS infrastructure completely reengineers the assignment process, making use of the inherent design features of the HFC access system, i.e., the TR303[1] or V5[2] interface, and the new capabilities of the network elements. Reengineering begins with the design of business-driven operational scenarios that detail how the provisioning process should work in order to meet service objectives of both the service provider and the customer. These scenarios detail the information flows, the OSS functional modules, the network element functionality, and the process steps required. Scenarios are reviewed and revised to accommodate particular business objectives and practices and become the basis for many requirements on both the OSS and the access system.

Table 7.1 shows an example of a simplified POTS provisioning scenario in a broadband environment. The net result of this process is a substantial reduction in the amount of data stored by the OSS. The shortened time required for provisioning could allow for real-time provisioning while the customer speaks with the service representative.

This ability is part of a concept called *customer care* in the telephone industry. The set of functions with which the customer has contact—billing, customer satisfaction management, and provisioning—were partitioned from the details of managing the network. A combination of OSS addresses those services.

Finally, the reduction of errors and manual work-arounds result in real cost

---

Table 7.1. Simplified POTS Provisioning Scenario in a Broadband Environment

Preconditions and required input data	• Resource provisioning has been completed
	• Customer address, associated HDT and NIU physical port
	• Available telephone numbers
Process	• The customer calls requesting service at a specified address
	• The address is used to look up the NIU and HDT physical port in the data base
	• The OSS queries the HDT for availability of the requested service and a spare CRV
	• The HDT responds with the service availability and a spare CRV
	• The OSS then formats the CRV into the Line Equipment format for the particular switch, and sends it with the selected telephone number and line features to the switch as the recent change message
	• The OSS sends a cross-connect message to the HDT indicating the CRV/physical port association
	• The switch sends a provisioning message to the HDT via the TR303 Embedded Operations Channel (EOC) with the switch line features
	• The HDT sends a service confirmation message to the OSS
Postcondition	The customer's service is provisioned and activated

savings to the service provider. Additional cost savings also result from simplified access architecture and simplified OSS architecture.

The Table 7.1 scenario also includes an example of the network element as the steward of certain data. Notice that the Call Reference Values (CRVs) are stored in the HDT and not in an external database. Because these data are actually used by the HDT to process calls, they must be correct. There is no reliance on external data to make assignments and produce errors that require manual work.

7.1.3. Software Architecture and System Integrity

When systems share a common architecture, they are the same. Architecture is the body of instructions, written in a specific coding language, that controls the structure and interactions of the system modules. The properties of capacity, throughput, consistency, and module compatibility are fixed at the architectural level.

The processing architecture is code that governs how the processing modules work together to do the system functions. The communication architecture is code that governs the interactions of the processing modules with data and with other systems. The data architecture is code that controls how the data files are structured, filled with data, and accessed.

Once the architecture is established, functions may be assigned to processing modules, and the system may be built. Processing modules can vary greatly in size and scope, depending on the function each performs, and the same module may differ across installations. In every case, however, the processing architecture, communication architecture, and data architecture constitute the software architecture that is the system's unchanging "fingerprint."

When several sites use software systems with a common architecture, they are considered to be using the same software system even though they may do somewhat different things. Alternatively, two systems with differing architectures can perform the same function although they do not do it the same way. They would be different systems.

For example, in the late 1980s, Bell Laboratories needed to develop a system to control a very critical congestion situation. The system was called NEMOS and had a *distributed* database architecture. It soon became apparent that the design problem was extremely complex and broke new theoretical ground. As there was no history of similar development for a guide and the need was urgent, Bell Laboratories decided, for insurance, to develop a second system in parallel. It was also called NEMOS, but used instead an *integrated* database architecture. The result was two systems with the same name, performing the same function. The system with the distributed database architecture failed, and the system with the integrated database architecture succeeded. They were two different systems.

There is a relatively new Chinese proverb that says, "A person with one clock knows what time it is; a person with two clocks is never sure." One might say the same of databases. For this reason, there is some wisdom to seeking software architecture with a single integrated database.

No two iterations of a software system are exactly the same, despite their shared architecture. When a system is installed at two or more sites, localization is always required. Tables are populated with data to configure the software to meet the needs of specific customer sites. The customer may have special needs that require more than minor table adjustments. Customization of some modules may be required. New modules may be added. Ongoing management of this kaleidoscope of systems is a major effort.

7.2. OBJECT-ORIENTED OPPORTUNITIES

Why do some systems succeed while other systems fail? This section will examine several case histories and isolate common themes that lead to success. Object-oriented technology can help to fulfill early computer industry aspirations and lead to predictable system developments with fast time to market and solid performance.

7.2.1. Thorough Business Understanding

Let's look at the story of the Lyons computer. The Lyons catering company in England supplied small tea shops with cakes and sandwiches in the 1950s. They ran their business with crews of women who would call each customer every day to check their current sales and calculate their needs for the next day. The farsighted management became convinced that the power of computers could be harnessed to increase their profits and serve their customers better. They actually built their own computer and succeeded in deploying an enterprisewide automation system. The changes ranged from anticipating orders based on past patterns to defining the order of loading the trucks for maximum efficiency in delivery. Soon everyone in the United Kingdom was eating Lyons cakes and sandwiches. This heady success led to the spin-off of a computer company that eventually became ICL.

The history of Lyons is noteworthy because it is the exception. Most software projects fail to live up to expectations. Often developers understand neither the business practices they are automating nor the business changes they will precipitate. Developers have also been known to do a poor job of producing software. Lyons's developers started from scratch. They controlled the requirements, and they worked closely with the hardware developers to build a system suitable for them. Later, force-fitting this system to others proved problematical.

Obviously very few businesses today present an opportunity for such a totally clean start, nor would it be desirable to refuse to build on progress. But the fact remains that interoperability is a problem.

7.2.2. Standardized Business Practices

Landauer[3] points out that by the early 1990s, with the exception of the telephone industry, computers had not led to measurable productivity gains. Telephone companies achieved a 2% per year productivity gain with the introduction of computer technology, a gain not seen in other industries. Telephone companies have spent a significant amount of their annual capital investment on information technology products and services since the mid-1960s. This long-term commitment has paid off. Today, information technology is at the core of the telephone business.

The telephone companies took a giant step to ensure success of their overall program when they invested in standardizing their businesses processes. They produced the Bell System Standards that defined how telephone companies would operate. As the business changed, these practices were updated. After the 1984 breakup of the Bell System, Bellcore maintained them. These business practices became the basis for the subsequent automation.

Of course, not all of the developed systems worked. Some missed the mark, but many served to increase the productivity of the telephone worker. Because there was a carefully maintained set of practices, there was ample opportunity to redo the systems that failed. With this habit of designing business practices first, it was natural for the telephone industry to quickly adopt object-oriented technology. Enterprise-wide object classes were derived from the practices. These object classes speeded the introduction of object-oriented technology.

7.2.3. Data Consistency

The case history of one modern telephone software system may be enlightening. The project was to support the use of new, very fast broadband networks in the telephone company plant. As this was clearly a large-scale development effort, the designers adopted the use of objects very early. The size of the project in its first release was 12,600 function points, 22 software modules with 47 interfaces, and 12 databases. This complexity was organized into 278 object classes and 1200 objects. The developers adhered to four overarching principles in making their design decisions:

1.   System synthesis, the melding of methods and business objects, began from the customer's, not the developers', viewpoint.

2.   Modular architecture separated data from applications and enforced strong data stewards.

3.   Object-oriented analysis included extensive domain analysis, rigorous requirements, business usage scenarios worked out with the user, formal external and internal interface agreements, and an integrated data model.

4.   Object-oriented design used client/server architecture and industrywide TMN standards.

The most serious problem on this project, which may be extrapolated to most business situations, was the need to keep data consistent. All of the older design methods used convoluted error paths to do this; these use more code and time than needed for building new systems, adding features to existing systems, or building bridges to legacy systems. Here object-oriented technology was a powerful tool for allowing quick system updates to accommodate new features and changes in business practices. It made reuse natural and forced system design to more accurately reflect the business objective.

It might be noted that until object-oriented design becomes a habit, an enforced object encapsulation strategy with centralized object libraries is vital. Skilled project managers must insist that all subsystems and modules use the same Operation, Administration, and Management (OA&M) software. This achieves meaningful reuse and results in huge system cost savings in operation of the system itself.

With these state-of-the-art object-oriented approaches, the developers on this project delivered in 18 months what every analysis showed should take 36 to 42 months. Unfortunately, this success was stillborn. The entire project was mothballed when the vagaries of real life intruded. It was impossible to obtain the rights of way to dig up streets and private properties to lay the broadband cable.

7.2.4. Experiences with Objects

That these results can be attributed to a disciplined use of object-oriented technology is corroborated by the experience of others. Swiss Bank Corporation[4] saw a 50% productivity improvement during reengineering efforts that started in 1991. By 1994, they were installing their new object-oriented system and said that reuse was the key to their success. The benefits of prototyping and adherence to clean object class definitions were particularly apparent. They managed risks by adhering to the standard enterprise object classes and linking them together. They anticipated some performance problems and these did occur, but the cost/performance improvement of new computer servers more than compensated for the 10% performance overrun they saw.[5]

Foster Wheeler reports that they can drive the applications building process by using objects with business rules. A decision-making process is modeled, then iteratively modified with time and experience. Rule bases are made part of the object methods so that the rules can be applied dynamically. This approach allows the inheritance and distribution of intelligence among objects at various levels. As changes are made to the rules, they naturally migrate to the affected objects. They saw that by using this approach, the time for projects was reduced consistently from 10 to 12 months to 6 to 8 weeks.[6] This may herald the reawakening of the expert system technology that held so much promise in the 1980s.

AT&T developed more than 50 object-oriented systems using a unique *objects in memory* approach.[7] The objects were locked in memory while the system ran. One such system may be the biggest and fastest object-oriented network management system in the world. It uses 1 gigabyte of memory for its 15 million objects and thousands of transactions per second on an HP high-end workstation. It has been in production for 3 years with no significant problems. It replaced a vintage IBM-hosted facility provisioning system. This new approach can become widely used when logical memory is extended to 64-bit addressing and added to the natural structure of object-oriented databases like Versant. This will open virtual memory machines to objects and regain freedom from memory constraints enjoyed by application developers in the earlier transaction systems.

The MCI Data Warehouse Project relied on off-the-shelf relational databases. It is a textbook example of the use of gateway and client/server relational databases. They used this technology to gather information for report production from many databases. Their multiplatform distributed set of databases consisted of IBM DB/2 as well as others. It had the look and feel of a single SQL server. This was not a trivial project. The challenge was to analyze, organize, normalize, link, and migrate data onto a database that end users could easily access without having to formulate complex SQL queries or write code. This is quite a different problem from developing databases that meet the stringent performance needs of network management systems. They routinely download data from network management systems to populate their data mining server. Here is a situation where object-oriented databases can live with relational ones. The performance needs are met with object-oriented databases while the flexibility for inquiries is met with a relational database.

In the MCI case, data analysts were called on to model data in a way to maximize usefulness to service planners. This meant rethinking the data model so that data availability and flexibility were maximized while retaining sensitivity to long inquiries. This was a hard data modeling job because of the several layers of indirection required to use the relational database management systems in the source systems. Direct use of object modeling and object-oriented databases, instead of relying of relational databases in most of the source systems, could have simplified this task and made the data models more flexible for unanticipated use. Without MCI's earlier and large investment in data management, this project could have easily failed, as research into a complex web of poorly modeled and documented databases is nearly impossible.

7.2.5. Objects in Large-Scale Projects

Large-scale evolving software presents a special challenge to object architects. Typically, an application consists of a network of objects connected through compatible interfaces. The need to meet new requirements and/or fix defects often results in new interfaces and object versions. When a new version of an object is created, it must be dynamically installed without causing disruption to existing software. Objects must be intelligent enough to handle the problems of dynamic reconfiguration, coordinate intermodule communication, and track the internal states of both the objects and the links. This increases the complexity of objects and can prevent them from being reused in different contexts. One solution is to not allow interface changes. This harsh rule often makes the application difficult to build because application-level interfaces are imprecise as a result of time-outs and repeated transmissions triggered by buffer losses in asynchronous communication. Additionally, the interface specifications are vague and not amenable to analysis.

In this dynamic environment, however, there is a premium for keeping all of the modules consistent. It is very difficult for designers, who are focused on the function of each module, to worry about the way all of the pieces will fit together. As a result, the issue of interface consistency is often left to test teams, where it is inefficient and time consuming. Experience shows that it is three times more expensive for testers to find and fix problems than developers. So, the interfaces must change but in a controlled way.

Object-oriented technology opens the door to dynamic checking of interface states and internal consistency because for the first time it is possible for projects to create libraries of interface object classes to do this job.

International standards bodies recognized this problem and developed the Common Object Request Broker Architecture (CORBA) standard to do distributed computing. CORBA is in its infancy, but industry cooperation is making CORBA the object middleware standard,[8] even though it has not yet provided the tools and methods needed for large applications. One deficiency is that it locks the sender until the receiver receives and acknowledges the message, and CORBA does not support multicycle transactions. CORBA's object module is evolving and may become the standard of choice.[9] The hope is that the object-oriented CORBA will provide the fabric to let architects connect their independently designed components together. OpenCon Systems Inc. and Expersoft Corp. are teaming to build a CORBA-based TMN gateway. Versant Object Technology has a powerful way of combining Orbix from Iona Technologies and their object database. The Versant/Orbix mediator adds the dimension of a database storing the objects used in the communication interface. Network Programs, Inc. provides the multiple transaction capability for mapping applications to one another. Their adapter/collector technology is a robust way of connecting systems while avoiding undesired interactions. Meanwhile, Microsoft offers its own brand-specific object approach, Distributed Common Object Model (DCOM), which allows clients to access servers across a network and provides a binary interface for packaging and linking libraries or other applications. It is not yet complete and may be linked to CORBA but might eventually have better supporting tools than CORBA. As the situation is still fluid, most organizations are using both approaches in combination with in-house controls for their interface designs.

Java applets and CORBA are well suited to building distributed Web applications. Browsers give access to network management data, and they allow networks to be managed remotely. To overcome delays related to network latency in the time it takes for a command to get to a network element, the *management by delegation* approach has become popular.[10] Data are stored in management information databases close to the network elements. They may be sent to the network elements as remote agents with their programs or they may be mapped to CORBA objects which can be located anywhere but must be statically mapped to the network element. Using Java applets, designers can overcome this limitation and dynamically reconfigure programs and their data objects. Java has a remote method invocation feature that is similar to CORBA but is restricted to only Java objects. CORBA can accommodate legacy systems; therefore, it is the best choice for distributing the network management data.

7.2.6. Summary

Object-oriented opportunities provide a way to realize the functional architecture described in Fig. 7.1. Customer Care modules, network management modules, and element management modules can all benefit from object-oriented techniques. This technology is finally sufficiently robust for large business applications.

7.3. MOVING TO OBJECT TECHNOLOGY

How can a company get on the object technology bandwagon? Most organizations either get everyone together and proclaim that they will now become an object-oriented shop, or they select a demonstration project in the hope that its success will spill over to the rest of the organization. Neither of these approaches is effective or efficient.

A very successful approach is to reinvent the business practices using object-oriented design techniques. This requires that object-oriented technology be at the heart of the enterprise's architecture. In this approach, all development organizations are funded to the extent that they conform to the new, overarching object architecture. With middleware being widely adopted for building robust enterprisewide applications, and with companies moving to thin clients (or at least thick ones controlled from a central system management organization), the time is right to embrace objects. Object technology ties the desktop, the Web, legacy, and client/server applications into a coherent whole.

7.3.1. Tools for Change

With the wildfire spread of Java and the adoption of component technologies, applications are being developed with extraordinary speed. New systems will likely be built around object libraries with Java to provide the way for running distributed architecture, as Java already integrates dispersed applications with remarkable ease on the Web. By coupling an object approach with rapid prototyping, software shops can reduce development time from years to months. The formal adoption of Boehm's spiral approach,[11] with its focus on getting the requirements right early in the development cycle, adds another tool for rolling out applications quickly.

The use of *jump-starts* has proven remarkably effective. One or two expert object-oriented designers, either hired from the outside or selected from within a project and intensively trained, are assigned to the development group for 3 months. They answer all of those nasty questions that developers struggle with during the transition from procedural programming to object programming. All parts of a system do not need to be converted simultaneously to object technology. A good approach is to start with the system administration and human interface functions, then move on to the core objects that model the business objective.

7.3.2. Changing the Organization

In the authors' experience with moving several organizations to object-oriented technology, there are certain common tendencies that are independent of application. At the initial introduction of object-oriented design, there is no gain in productivity in the first release. Great energy is invested in training, building a software factory, building tools, making sure people understand how things work together, and choosing people who can make the transition from ordinary procedural work.

Even under these circumstances, however, there are benefits right from the beginning. Modifications and extensions to the first system release are two to four times faster than by previous methods. These results reflect having code that is free of structural defects. Though functional problems may remain, the code is easier to test, and therefore problems are isolated faster. In addition, the software modules are largely reusable, which saves on major recoding efforts.

Not everyone is suited to object-oriented design. Restricting the number of object architects to approximately 10% of all programmers/designers and isolating the remaining 90% to do normal procedural programming seemed to work well. Everyone was required to use object libraries for interfacing between processes. Experts in C++ and object-oriented design were critical for training other engineers. They functioned as industrial engineers do to prevent the misuse of a tool; they are available during the process to help the team design and build the software within the constraints of the tool while exploiting its full potential.

Software middle managers, who are generally risk-averse, were appointed only after the architecture existed and the first iteration of development had begun. Prior to that, the team was isolated and protected from interference and discouragement while they laid out the technical strategy.

Developers who went through such an experience felt pride in personal growth, recognized that object-oriented design made their maintenance job easier, and expressed a sense of high morale and camaraderie with their peers. If the same team goes on to use the techniques on subsequent projects, they show two to three times the productivity over procedural programming.

Certain constraints are useful. The architecture team is encouraged to build a throwaway prototype of the object libraries in order to gain experience with objects and with mapping into the problem domain. Design rules concerning memory, language constructs, and communications through object libraries must be observed to get the design right. The advantage of doing this well was that problems were readily isolated. Care has to be taken in nesting object classes in terms of inheritance, or the software build time will grow exponentially; object architects need to be careful not only about execution, but also about build sequences for testing and production work. All design can be simplified on the second pass. The team should be able to reduce the number of object classes by 20% and similarly reduce the number of function points after the prototype. Templates for object classes and limiting the number of global objects worked well.

While developers will applaud the progressive thinking of the executive team that institutes object-oriented technology, middle mangers resist. Middle managers see the need to retrain themselves and their people while they try to meet tight schedules and cost goals. They are not confident that they can technically manage an organization using unfamiliar new technology. For that reason, a gradual evolutionary approach is best, but the point must be made that object-oriented technology is inevitable and desirable. Software shops must adopt this technology or become obsolete.

7.3.3. Summary

Object databases are mature; they crack performance problems and support the use of enterprisewide object classes. A rich integration of object and relational databases with a strong focus on module interactions is today's best current practice. Organizations adopting object-oriented technology will gain a threefold productivity advantage over those that remain with procedural methods. This is today's need—on tomorrow's horizon is transparent computing using Java interpretive programming.[12] This technology holds out the promise of another threefold increase in productivity and regular achievement of 80% reuse.

7.4. CONFIGURATION MANAGEMENT

One story will serve as an example of the unanticipated consequences of inadequate configuration design. It is from the authors' experience and illuminates the pitfall of listening to the customer's *wants* without fully understanding the customer's *needs*.

7.4.1. You Want WHAT?

In 1985 when client/server systems were young, we did not know we were actually doing data management. The problem we faced was to broadcast work status information from a Unisys mainframe to 30 NCR tower clients throughout the day. A work center manager who wanted to know how much work was left could ask the local client. This was a nice feature most of the day, but became critical at 3:00 PM when all of the managers wanted to know if they had to schedule overtime to close out the day.

Before we had the client/server solution, they would jam the mainframe and networks with report requests, generations, and transmittals. And each wanted only that work that was related to their technicians. To keep the clients in step with the mainframe server, we provided an initial report whenever the client came online. This obviated the need for a separate record of the state of each client, because we relied on the clients to customize the one comprehensive report for each work center. The server would broadcast all changes to all clients. By using this approach, we did not have to resort to complex startup and recovery procedures that cost network and server capacity.

Today, better solutions elude us except for client/server systems that do not grow too fast in size or capability. Static mapping of data models or of software executables will not be good enough to handle future applications. Who will be charged with keeping this complex of systems operating sanely?

7.4.2. The Network Is the Database

Autodiscovery of clients on a TCP/IP network has been wonderful, but this feature does not scale well and has been used sparingly in telephony applications. A generalized autodiscovery feature is needed that embraces the concept that the network is the database. It is also needed at the services level. Recently, Hewlett-Packard added centralized software to its OpenView platform that manages configurations across networked systems. Instead of making multiple changes every time a user is added to the network, the administrator can use one command to configure the user's password, e-mail account, and downloaded software. HP relies on a synchronization function to reconcile the actual state of the networked applications and computers with the administrator's databases. The problem of mixing the network's physical inventory with logical data in UNIX databases is formidable.

One detail will illustrate the scope of the problem: When we built a prototype to extract information from a network element and write it to a relational database, the hardest part was getting the client protocol stack just right, especially in its interaction with the server's relational database. The database demanded versions of the protocol stack that could not be purchased for the client. Once a specific configuration worked, it worked well. Change, however, is problematical because the environment is not robust to changes in the client or the server.

7.5. BUGGY SOFTWARE

Software may be the toughest problem to solve in building systems that manage other systems. Software has the awful propensity to fail with no warning. One manager of our acquaintance issued a memo stating, "There will be no more software bugs!" The trouble was he meant it—no joke. Even after bugs are found and fixed, how can software be restored to a known state, one where its operation has been totally tested? For most systems, this is impossible except with much custom design that is itself error-prone.

One new idea is software rejuvenation. It is special software that gracefully terminates an application, and immediately restarts it at a known, clean, internal state. Instead of running for a year, with all of the mysteries that untried time expanses can harbor, a system is run for one day, 364 times. It is reinitialized each day, process by process, while the system continues to operate. Rejuvenation precedes failure, anticipates it, and avoids it. It transforms nonstationary, random processes into stationary ones.

Increasing the rejuvenation rate reduces the cost of downtime. Two years of operation has passed with no reported outages for one system set at a weekly rejuvenation interval. In another laboratory, a 16,000-line C program with notoriously leaky memory failed after 52 iterations. After adding seven lines of rejuvenation code with the period set at 15 iterations, the program ran flawlessly.

Rejuvenation does not remove bugs that exist beyond its carefully circumscribed limits. Instead, it avoids the vast unknown territory that conceals them.

7.6. MORALS OF THE STORIES

The collective wisdom so far concerning the most critical functions and features of software are as follows:

- Designing the OSS and intelligent network elements together makes operations easy.

- Success depends on capital investment and standardization of business practices.

- Object-oriented design standardizes interfaces, allows an efficient redesign of business processes, and through these means increases productivity.

- *Data*, not *database*, management best describes the job in complex systems and networks.

- Software is conditionally stable. Large, complex systems cannot be totally tested, so software rejuvenation is a technique that can regularize software execution until a fundamental theory of software reliability can be developed.

REFERENCES

1. Mulla, H. D. "HFC network applications and design issues," *Annual Review of Communications, International Engineering Consortium* **50**, 527–531 (1997).
2. Gillespie, A. *Access networks: Technology and V5 interfacing* (Artech House, Boston, 1997).
3. Landauer, T. K. *The trouble with computers: Usefulness, usability, and productivity* (MIT Press, Cambridge, MA, 1996), pp. 13–35.
4. Personal communication with L. Bernstein.
5. Graham, I. "Making progress in metrics: Task-point analysis can be performed at the requirements stage," *Object Magazine* **6**:8, 68–73 (Oct. 1996).
6. Vaughn, N. "Corporate success stories, integrating objects with rules," *Object Magazine* **7**:1, 66 (March 1997).
7. Bergholm, J. O., Davis, J. M., Nadji, B., and Ting, P. D. "Service design and inventory system—An object-oriented reusable software asset," *AT&T Technical Journal* **75**:2, 47–57 (March/April 1996).
8. Gaudin, S. "Object stamp of approval," *ComputerWorld* **31**:11, p. 1 (March 17, 1997).
9. Rixon, J. "ATM management using HP DM CORBA and Java," *CiTR Technical Journal* **2**, 47–56 (1997).
10. Yemini, Y., Goldszmidt, G., and Yemini, S. "Network management by delegation," *Integrated network management II* (Krishnan, I., and Zimmer, W., eds.) (Elsevier, Amsterdam, 1991), pp. 95–107.
11. Boehm, B., and Bels, F. "Applying process programming to the spiral model," *Software risk management* (IEEE Computer Society, New York), pp. 38–46.
12. Bernstein, L., and Yuhas, C. M. "And the walls come tumblin' down," *IEEE Communications Magazine* **30**:12, 126–133 (Dec, 1992).

Pillar—Database Considerations

"Just the facts, ma'm," Jack Webb would say, as if it were so simple. Having the facts is critical to having effective broadband network management systems. Data must be reliable, their storage cost effective, and their retrieval and maintenance easy.

8.1. MAJOR DATA QUESTIONS

It has been a difficult task to keep data that, first, describe network elements, second, show how they are connected together, and third, prescribe when and why these connections will change. A key issue implicit in these specifications is the need to automatically trigger the work needed to install a new telecommunications service or recover from an outage. The huge size of the networks under consideration and the number and complexity of the problems that can afflict them make human control of scheduling and plant dispatch untenable.

8.1.1. Non Confundar in Aeternum*

Computers come and go, software gets written and overwritten, but data are eternal. Much of the data used today were put into computer storage in the 1970s and 1980s. There have been attempts to build systems that were driven off a common database or at least a single data model, but these attempts failed because the technology was simply not up to handling the complexity of telephony data models, nor were computers fast enough to supply the work center needs.

As a second best effort, separate systems grew up to support individual work centers with local versions of data. The local version will, of course, be what is needed for the immediate situation, with no consideration for standardization to some total system solution that may or may not ever come.

---

* Translation: Do not confound me for all eternity.

For example, the simple notion of storing someone's telephone number is not so simple. One system might need to track area code while another does not. Some computers serving a frame or a switch tracked only the last four digits of the number to gain maximum use of the expensive computers in the 1970s, because the first three digits, representing the switch, were common to all local work. What the data were called differed: One system might refer to telephone number as *TN*, while another might call it *Tel. No.*[†]

8.1.2. Data Naming

The ability to personalize data names was and continues to be important to system managers and application developers, perhaps reflecting a strong need for individual recognition in a large and impersonal effort. Provisions for using local synonyms for global names are very popular. Variation comes with a high price, however. Meanings differ. Structure differs. When systems try to communicate, there is trouble.

At least four parameters can be locally unique. When data are variable on so many parameters, there are worries sufficient to tax even the universal translator of Star Trek fame.

- *Syntax*, the bit structure, must be intelligible.

- *Semantics* is the meaning of data within their context—TN varies depending on where the call originates (e.g., an office with a PBX, a car phone, a residence in the same area code as the call termination).

- *Mediation* is the question of doing data updates on differing schedules and having to resolve differences.

- *Data exchange rules* negotiate various systems' expectations for sending and receiving. These are combined into various protocols used to connect systems to one another.

[†] Carroll, Lewis. *Through the Looking-Glass and What Alice Saw There*, 1872.

"You are sad," the Knight said in an anxious tone: "Let me sing you a song to comfort you."

"Is it very long?" Alice asked, for she had heard a good deal of poetry that day.

"It's long," said the Knight, "but it's very, *very* beautiful. Everybody that hears me sing it—it brings the *tears* into their eyes. . . . The name of the song is called '*Haddocks' Eyes.*'"

"Oh, that's the name of the song, is it?" Alice said, trying to feel interested.

"No, you don't understand," the Knight said, looking a little vexed. "That's what the name is *called.* The name really is '*The Aged, Aged Man.*'"

"Then I ought to have said 'That's what the *song* is called?'" Alice corrected herself.

"No, you oughtn't: that's quite another thing! The *song* is called '*Ways And Means*': but that's only what it's *called*, you know!"

"Well, what *is* the song, then?" said Alice, who was by this time completely bewildered.

"I was coming to that," the Knight said. "The song really is '*A-sitting On A Gate*': and the tune's my own invention." . . .

"But the tune *isn't* his own invention," Alice said to herself: "it's '*I Give Thee All, I Can No More.*'" She stood and listened very attentively, but no tears came into her eyes.

There were enough problems and enough isolation of systems to provoke some rethinking of the data storage and access issue.

8.1.3. Attempted Solutions

The first thought was "family style service"—one pot and everybody dips in. Instead of being repeated in many databases, data could be stored in a common database available to all applications. This approach requires a great deal of system engineering to make sure that all work centers can be served well and that changes needed for one application do not impact others. There must also be gateways into and from the common database that are recognized by each application.

Another approach is to store data in several places but to rely on a single source for accuracy and structure, using a common data dictionary. In either case, great care must be taken to make sure that the stored data accurately reflect the state of the network. This is the problem of data *accuracy*, and much effort is spent checking the actual network elements against the data. When it is not possible to have a common data definition, the problem of keeping data *consistent* among systems arises. Multiple databases with multiple versions have driven technicians to trust only the notes they can carry on scraps of paper in their pockets or what they can see with their own eyes in the field. The irony was not lost on many people.

Figure 8.1 shows how a set of data used for traditional loop provisioning

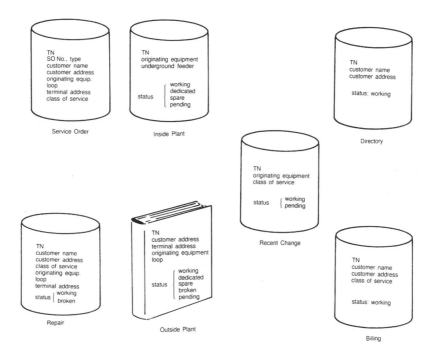

Figure 8.1. Multiple databases.

evolved into many systems. At first, the original paper records were used to populate several databases with different combinations of the data. Later, a database used by one work center might be used as a base for new systems. These databases could not be kept in lockstep because they each had a different updating approach. Even the best circumstances created consistency for only a moment. Just before installing a new system, a substantial data bashing effort establishes databases that are somewhat close. For example, trouble-ticketing databases can be out of step with assignment databases, and they both often lag behind the customer service databases.

8.1.4. Reasons for Failure

Few attempts to have an integrated database were successful because the technology of the 1980s could not easily scale to the size needed for all of the work centers, and too many performance bottlenecks hindered the workflow. Figure 8.2 shows how a few telephone companies have developed an integrated model for small and medium-size COs. These database designs cannot scale to manage COs of 75,000 working loops or larger.

A model of the loop plant was developed to understand the relationship between the data records and the important physical and administrative characteristics of the network elements. Later these data records became objects when the *methods* for processing the data variables were added to the records.

For example, a pair used in a loop can have its status, its physical composition, and its connections stored in the data records. A loop plant database is partitioned into CO modules requiring 100,000 to 5 million data records to store all of the information for 10,000 to 100,000 pairs.[1] Because of the number of transactions needed

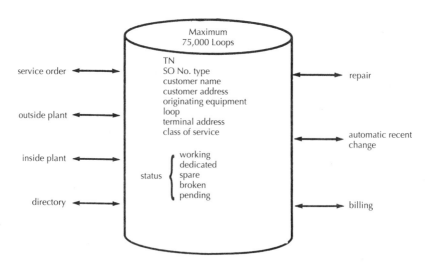

Figure 8.2. Common database.

to process all of the changes made to these loops in a typical workday, sophisticated database locking techniques with the ability to run several transactions in parallel are used. Even with these techniques, it is often faster to physically change the configuration of the physical plant than it is to change the databases to reflect that change.

8.2. WHERE DID ALL THOSE ERRORS COME FROM?

There is a correlation between customer satisfaction and the major source of error in communications equipment, which is plant errors induced when craft technicians change configurations, repair known problems, and troubleshoot intermittent problems. In the worst case, errors can propagate in a ratio of 1:3 to 3:1.

8.2.1. Hypothetical Case

Let's consider Worst Service Telephone Company (WSTC). At WSTC, they are very good at minimizing capital investment by pushing the utilization of plant facilities toward 95%. Unfortunately, when a request for a new line comes in, that high utilization means there is rarely a spare where it is needed. The craft technicians have to move two existing lines to make room for the new line. On average, one new line equals three plant changes. That is the 1:3 part of the relationship.

WSTC also has a problem with craft-induced errors. Management is very thrifty about the number of craft they hire, and they push hard to get the maximum number of jobs completed each day. They do not invest time in training. They do not spend capital investment on new technology. It is no surprise then that craft introduce one fault into the plant for every three rearrangements. That is the second part of the relationship: three rearrangements equal one error, or 3:1.

This leads to perpetual work. The more the company grows, the more work is required to fix problems. Even without line growth, the 1:3 to 3:1 dynamic ensures continuing work for WSTC craft. Even at the Pretty Good Telephone Company (PGTC), the error introduction relationship is 1:1.5 to 5:1. No one escapes this reality of human intervention.

8.2.2. Empirical Evidence

Many telephone companies have noted a remarkable decrease in repair problems during strikes. Installations often are deferred during strikes, and there are fewer people making changes to the plant. When changes are needed, the first line managers who are working the craft jobs take special care not to introduce problems. When they see a potential problem, they stop and fix it (taking a certain pride, perhaps, that they haven't lost their touch). They have the luxury of seeing the whole purpose of the activity.

The phenomenon that equipment is noticeably more reliable when it is "buttoned-up" is well known to network managers. Military test directors know that they must freeze a system well before it is in use. But one cannot freeze communication configurations when customers are demanding new services.

The need to rearrange plant is not related solely to craft practices and line growth. Plant deterioration and the demand for new services also cause substantial plant rearrangements. Taken together, these rearrangements may be described as plant *churn*. Studies have shown that as little as a 5% churn rate leads to a 25% increase in provisioning and maintenance expenses in our theoretical WSTC. Even for the PGTC, a similar churn rate leads to a 20% increase in provisioning and maintenance expense. This expense growth argues strongly for investments to reduce the need for crisis-driven plant changes.

8.3. DESIGN ISSUES

Getting the data model right is critical to having a good network management system. The problem is that several models of network management systems have gained in popularity. For example, SNMP has its own data model and language; CMIP has a GDMO model. When used together, they are incompatible, and it is necessary to do transformations to allow communication. A new common information model is being defined to try to deal with differences in an evolving standard. Similarly, a TINA-C Network Resource Information Model is being introduced, which is a way to inject object-oriented technology into the process. Middleware also imposes constraints, be it CORBA or Microsoft's DCOM, or any other model. Therefore, the designer must deal with constraints of both the problem domain and the software model.

The new Unified Modeling Language (UML)[2] offers the hope that a high-level design technique will prove useful. The language facilitates the definition of object classes and makes it convenient to program attributes of network elements within the object classes. This approach is a natural extension of earlier object design methods. It is important to make sure that these attributes be fundamental characteristics of the network elements and not derivable from other attributes. Modeling derivable attributes as independent ones often leads to mismatches between objects and flaws in the software design as well as errors in the database.

8.3.1. View of the Assets

As telephone companies shift from a functional organization to one concentrating on providing end-to-end services, they find that the designs of their OSSs are not well suited to a broad-based view of their assets. To improve customer service, they must design online ordering, trouble management, billing inquiries, and customer care on top of an integrated data model. The data model integrates customer service records with service databases and equipment databases. When actions are needed to change the network elements, a work force management application is triggered. As the work is finished, the application tracks confirmations. That tracking and data capture were not automated created a situation that left databases open to error.[3] It becomes more important that tracking systems treat service provisioning end to end and coordinate with billing systems when broadband networks are self-configuring and customers can create their own services.

8.3.2. Management by Delegation

The powerful approach of management by delegation, which was discussed in Chapter 7, allows designers to place the details of triggering and coordination with an agent located within or close to the network element. A special manager–agent delegation model allows management programs to execute locally at the agents, calling on the manager only for high-level tasks. Relational and newer object-oriented databases with distributed processing make it possible to use this technique, which has the advantage of encouraging technicians to grasp the connectivity. Object-oriented databases have the additional advantage of more easily meeting timing constraints and eliminating a level of abstraction in the data record design.

8.3.3. Performance Demands

Broadband networks present special performance demands. A study of MIB implementations compared the use of a relational database, an object-oriented database, and MIBs that stay within the processor memory of network element controllers.[4] As expected, the memory resident MIBs were the fastest, but they presented serious maintenance problems. It is difficult to keep the MIB current with network element features, and storage was at a premium. Where computers can serve several network elements at a control point, the object-oriented-based MIB provides a good compromise among the factors of speed, storage, and maintainability.

8.3.4. Essential Design Features

Figure 8.3 shows the essential design features for a data management system. Transformation systems convert one data model to another to allow application software to gain access to the data. Sometimes this requires mapping one transaction to many transactions. Consider the example of a distributed database system going into an application that needs data from several databases. The transformation system will go to several database management systems, put the information together, then transform it from the syntax of the source into the syntax of the application software.

Defining the rules is part of the job of the transformation process to adapt databases to applications. One might call this general class of system *adapter/ collector* systems. The input and output schemes for databases should aim to reduce the need for adapter/collector systems as databases evolve in sophistication, because this is simply a troublesome boundary condition where adapter/collectors try to ameliorate uncoordinated design. Data dictionaries try to deal with the same problem, but they lack the scope to handle all of the data items of a large enter-

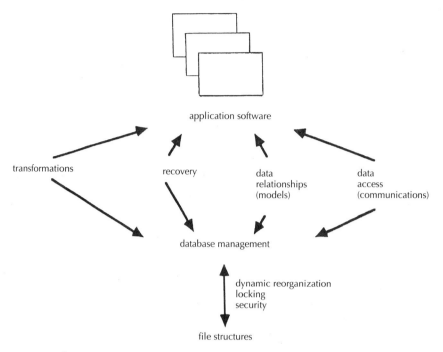

Figure 8.3. The data management system as a unifying concept.

prise. In the case of data dictionaries, initial success is possible with great diligence, but upkeep is prohibitively expensive.

To continue with Fig. 8.3, the transformations are contained in special communications objects. The recovery systems allow transaction replay. The data relationships contain the detailed object models for the network elements. The data access contains the protocols needed for distributed processing.

When client/server solutions were first used for network management, they were not tightly linked to the large host data servers. Individual departments in telephone companies installed their own OSSs operating on minicomputers to extract information from host databases but did not provide a way to keep the databases coordinated. The databases degraded further with the advent of the personal computer, which updated neither the minicomputer nor the host.

Figure 8.4 shows how data are abstracted from large database systems for various disciplines. Distribution plant and line equipment data, for example, may be used by engineering and marketing in one set of systems and by operations in another set of systems. Because there is no good feedback system, the occasional database "bash" must be done to resolve differences, but this is a costly and time-consuming process.

This design attempted to gain control over data consistency, but it was ill-conceived from the start because it took no notice of how people actually work. It removed from the work centers the responsibility for their own data and transferred

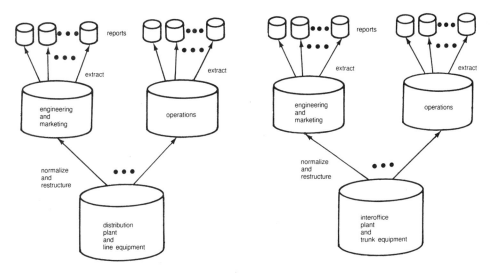

Figure 8.4. Hierarchy of OSS databases.

that responsibility to system administrators who could not understand the needs of the line workers. One example of many will suffice. When this design was common, there were serious backlogs for the creation of report generation systems. Line workers, however, needed to produce critical reports for filing with regulatory commissions. Therefore, they would use their PCs to download what they needed for their reports and neglect to make changes back to the corporate mainframe. System administrators, in a misguided attempt to conserve CPU cycles, then turned off access so that the work centers could not do this. Figure 8.4 shows the data migration from the central database to PCs without any systematic provision for keeping the database consistent and accurate. Data stewardship was eventually and appropriately returned to the work centers as will be explained in Section 8.7.

8.4. WORLD WIDE WEB

Using the World Wide Web on the Internet is a logical evolution from the client/server solutions. The gradual deployment of ATM supports multimedia transmissions, and more attention is being paid to various QoS factors end to end. It is changing the way companies provide data to their employees and their customers.

The Web consists of interconnected servers giving data to clients. Most systems use relational databases, though some network management applications are moving to object-oriented databases from companies like Versant. These databases use a gateway, the Common Gateway Interface (CGI), to provide data needed by clients. Standard Query Language (SQL) selects data that match given criteria. Making the Web connection to network management databases follows these steps:

- A client machine running a Web browser requests information from a Web server.

- The Web server processes the request from the client and runs a CGI program that connects to the database.

- The database server runs the requested SQL program.

- The CGI program retrieves the result of the SQL inquiry and relays it to the client. Any needed special translations to display the data are done by the Web server.

This scenario is simple but shows the fundamental steps any integration of the Web with data servers must follow. Security and authorization must be handled properly to prevent abuse.

The Web is being used for network management and to provide customer access to customer care systems, allowing telephone company customers to deal directly with network management software. Web applications are being used to link network management systems into work centers, particularly as they become more distributed and as technicians become more mobile.

8.5. SETTING UP DATABASES

The need to manage new services through service control points can be demonstrated by the example of local number portability. When a customer changes local service providers and wants to keep the same telephone number, a service control point accesses a service management system database to route calls to the correct loops. This architecture builds off that used for toll-free 800 services. The difference is the large number of inquiries that the service management system and its supporting SS7 network (used for number routing) must handle. The SS7 network may be faced with twice the traffic volume. This can induce bottlenecks. In addition, billing systems must become independent of telephone numbers.

8.5.1. Build the Relationships

Special service design and installation systems build the relationships that are loaded into the service management system. When these systems are difficult to use and the relationships are hard to represent in databases, new services take precious time to implement. New service management systems have the goal of speeding up activation and packaging. End users are offered ways of constructing their own services, further complicating the database design. Once these systems are in place, it will become more difficult to manage SS7 networks.

The management problem was first seen in building databases for fiber-optic networks. In those instances where tools were put in place to build reliable and consistent databases, network management became possible. If the task of populating the database was left to technicians with no guidelines, trouble abounded.

8.5.2. Example

In the example of one case, the configuration was based on multiple transactions. Data entry technicians were able to put only simple linear relationships into the database. The complexity of building the database to show interactions between network elements and their hierarchy was so difficult that most of the system features were rendered useless. One feature that suffered was the ability to suppress alarms and avoid alarm avalanche should disaster befall a major fiber facility, such as being dug up by a backhoe. In contrast, when a CAD-like system containing rules for building both the network and the database according to service relationships was used to build the database, the network management system was able to radically improve the availability of the fiber-optic network.[5]

8.5.3. Warning Signs

Recent statistics showed a 35% error rate in local number routing databases; warnings were also sounded that designs must handle the 15-digit international dial plan.[6]

Any database configuration should have ways to modify the structure (and in the future some ability to alter schema) without bringing the system down, and the database manager should support the creation of classes at run time. Object-oriented databases go beyond the typical "alter table" command of relational databases.

8.6. DATA WAREHOUSING

Engineers can now design systems that integrate various data into models that show relationships, trends, and correlations. A Data Warehouse (see the discussion of the MCI Data Warehouse in Section 7.2.4) is a very large database server with connections to application OSSs and to a client/server network.[7] Data records are regularly taken from functional OSSs and processed for inclusion in a Data Warehouse. Not all data are under the stewardship of the Data Warehouse, so inconsistencies and inaccuracies may arise.[8] Data gathered from an appropriate OSS are consolidated with other data, and any differences are resolved to create an integrated data model. For example, telephone companies need detailed knowledge about how customers use services in order to target those who might be receptive to new service offers. A customer Data Warehouse containing 3 to 12 months of customer usage information can provide the necessary databases for tracing customer preferences, installed equipment, or billing data.

8.6.1. Quality of the Data

The reconciliation process is key to the quality of the data. Because these data are used to project the changing nature of the business, there is a premium on accuracy. Before Data Warehouses, inquiries against OSSs were done ad hoc as a special project. Much energy was spent to establish the validity of the results, and often inconsistencies in the OSSs were not fixed because the cost in time and effort was too great. With Data Warehouses, the situation is different. If a company has invested heavily in a working data dictionary, each data item has a data steward. When inconsistencies are found, the Data Warehouse software returns to the data steward for reconciliation. This reconciliation may take the form of inquiries directly to those network elements that maintain data records, or they may rely on one OSS.

The best strategy is to choose consistency over accuracy and use reconciliation software to update the OSSs that do not agree with the data steward, even in cases where the data steward has inaccurate data. Once the inaccuracy is noted in the steward and is corrected, the updating software naturally provides feedback to the other OSSs. This sometimes causes a ripple effect of data changes, but where network element data reflect the actual use of the network, the databases rapidly converge.

To gain this advantage for older network elements that do not keep data records, element management systems have been developed to encapsulate the network elements and maintain the critical data. Administrative data in customer care systems need to be regularly checked for data consistency so that the ripples do not become waves. Often data scrubbing software systems are run regularly to keep the data free of inconsistencies. When data scrubbers are tied either to systems that do a physical inventory of network elements or to the network elements them-selves, the data scrubbers are used to make the data accurate.

8.6.2. Alternative to Physical Data Warehouse

An alternative to having a physical Data Warehouse is to keep a logical dic-tionary with pointers to where information is stored. Collection, consolidation, and reconciliation of data from OSSs could then be used for business inquiries. This strategy has not worked in practice because gathering data can impact the opera-tional effectiveness of the OSSs, and real-time processing causes bottlenecks and long delays in getting results. It is better to smooth the work loads and use separate systems for business inquiries.

8.6.3. Constraints

The successful Data Warehouse observes the following constraints:

- The Data Warehouse is read-only to clients making inquiries.
- The Data Warehouse is not a data steward.
- The Data Warehouse contains data for decision support, not for routine operations.
- The Data Warehouse provides only consistent (not necessarily accurate) views of the data.
- The Data Warehouse is used mainly for enterprisewide data.

8.6.4. MIB Data

The MIBs for SNMP or CMIP must contain a current model of the network and accurate data. A conceptual database model is constructed and updated to reflect the current state of the network and planned future states.

There are two general approaches for collecting the MIB data, dynamic and static. The difference between them resides primarily in the question of how many copies of the data are kept.

In the *dynamic* approach, the agents send data after a network management application requests them. The MIB can be dynamically updated so data are timely, accurate, and consistent. The dynamic approach has the advantage of sending only needed data but has the disadvantage of not being able to retrospectively trace the conditions that caused an outage.

The *static* method collects all possible data, independent of its use, so that designers do not have to anticipate data needs. Databases are loaded at the time of network changes and updated separately from the operations system. This method is generally used for large systems for managing network configuration and updating tables. MIBs are consistent, but it results in lots of data.

Communication delays and component failures make it difficult to build an accurate and consistent snapshot of the state of the system. There are dozens of factors that can be measured to determine QoS, such as throughput, bandwidth, transit delay, residual error rate, establishment delay, and so on. For broadband networks, it is important to track such measurements in MIBs.[9]

8.7. TRACKING CHANGES TO THE DATABASE

The problem in Fig. 8.4, databases that were consistent but inaccurate because of poor human factors design, needed a solution that involved a lot of feedback in a constantly changing situation. Figure 8.5 illustrates a dynamic method for accomplishing this. The middleware developed in the 1990s allowed coordination and recovery of individual databases that led to better operation of distributed systems. The identification of data items from local work centers was permitted by using object-oriented technology for communication. Line people regained corporate data and kept them up to date. The large number of feedback paths illustrates the complexity of keeping data current.

8.7.1. Example: Install New Cable

One example will show the dimensions of the problem. When a telephone company decides to install new cables, termination equipment, connection equipment, or carrier equipment, they refer to their fundamental plan. This plan is based on projections provided by marketing organizations for line and service growth. The activity that specifies and coordinates the work of all of the telephone company people needed to make these changes is called a *cable throw*. The outside plant engineers refer to in-place equipment to determine the most economical upgrade. They

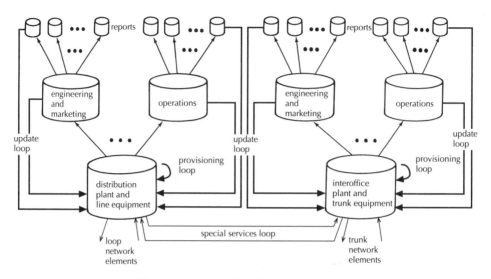

Figure 8.5. Dynamics of database change.

then select the best cable throw design and send it to a provisioning system for execution. The provisioning system is responsible for managing current logical assignment activity and producing the construction cut-sheets as is customary. After the construction work is complete, equipment can be assigned and activated as part of the service provisioning process.

If other changes are needed, a system must examine the physical composition of the new cable and pairs to detect changes inconsistent with the needs of the broadband service. A database tracks the current condition of the plant and all future changes. If suitable pairs cannot be found or if there is a pending change incompatible with the service availability date, the outside plant engineer selects a pair not yet installed for the service.

An application assigns the work to the needed technicians and tracks the completion of the construction work to ensure that the pair is ready when the customer needs it. If it is not, a notice alerting the outside plant engineer to an impending missed due date is generated.

To provide midband service, a database tracks changes to in-service loops and makes sure that they remain qualified to carry the service through any changes made to the loop. The information that the loop is midband qualified is kept in the logical extensions of the database. Qualification includes, but is not limited to, checking data records containing the loop makeup, the presence of load coils, the capacitance, and any test measurements taken on the loop. Construction work is carefully monitored to ensure that a qualified loop does not inadvertently become unqualified.

Data connections are made to the customer care systems so that the availability of lifeline service, midband service, POTS service, or satisfactory loop makeup can be checked. This preorder status check is possible, but not in the Data Warehouse. Special interfaces between the engineering, assignment, and customer care databases would be built for these routine checks.

8.7.2. Fully Automatic

Today this activity and others like it are done by human intervention. Clearly, the complexity demands that it become fully automatic in order to maintain accurate, consistent databases.

8.8. POTENTIAL DATA WAREHOUSE USE

Telephone companies have long needed a system that combines physical information about how their copper loop plant is built with logical information about how it is used, and geographical information about where it is installed. The cost of conversion has always outweighed the direct benefit of having such a database. Broadband services, however, are very sensitive to plant changes. A Data Warehouse can combine different outside plant data models to quickly and reliably compute the electrical characteristics, or *loop makeup*, needed for designing special services. It can also provide machine aids for the design of cable throws targeted at relieving congested feeder and distribution routes.

The physical maps of telephone poles, underground conduit, terminal boxes, and other telephone company equipment are kept on paper or in CAD systems. The logical assignment of equipment is kept in different systems. Some systems use extractions to tie the data models together, but these are not complete. As companies move to broadband networks, they capture the outside plant physical equipment in computer databases. Many are using a special Information + Graphics Systems data model, called DB-ABLE, for this job to store data in logical form while still being able to produce blueprint-quality engineering drawings. The benefits of such integration are widely known for the traditional copper plant. The introduction of midband capability like ADSL or VDSL makes the usefulness of such a database even more compelling.

8.9. CONCLUSIONS: CHANGING DATABASE HEURISTICS

The cable throw task force was responsible for resolving all of the problems involved in coordinating and meeting many departmental schedules. Large networks happen too fast and are too complicated for slow human meetings to be effective any longer. Automatic dispatch of field technicians, based on embedded logic in the databases to coordinate dependencies, is what the situation demands.

Databases have linear threshold information about when changes should occur; this is usually one parameter, like a date or an event. Databases that are date-driven are thwarted by life's little surprises. The same crews that are used for traffic peaks and schedules also respond to emergencies, which tend to invalidate schedules. Because it seems so logical to use dates as triggers, database designers do not think about the problem as one of multiple interdependencies. Requirements writers cannot capture the true complex nature of the work flow.

One resolution is available in object-oriented design. Object methods can contain expert system rules. Criteria can be established to integrate changes due to events and schedules. Engineering activity also happens over various cycles, day-to-day problems, weekly objectives, and semiannual fundamental planning. Object-oriented design can coordinate these.

New services change the database heuristics. Inheritance schemes make it possible to populate all object methods with the new rules. The *knowledge rules* base can move toward the *data*base now that the technology exists in object-oriented design to create expert systems.

OSS planners need to borrow a phrase from environmentalists and think globally while they act locally on the data in their systems. Capturing, reconciling, and making data available for technicians has been difficult. It will be even more challenging to make data available to customers and, because regulators now require it, to competitors. To have a successful data strategy, service providers will be obliged to capture data directly from network elements, use object-oriented technology to hide data complexity from users, and order plant changes from applications using logic embedded in the database.

REFERENCES

1. Goldstein, A. J. "A directed-hypergraph database: A model for the local loop telephone plant," *The Bell System Technical Journal* **61:**9(Part 2), 2529–2554 (Nov. 1982).
2. Melewski, D. "Connecting the dots," *Application Development Trends* **5:**2, 25–32 (Feb. 1998).
3. Ridgeway, J. "Automated business interconnection," *Telephony* Feb. 9, 1998, pp. 42–44.
4. Shimizu, T., Yoda, I., and Fugii, N. "Implementing and deploying MIB in ATM transport network operations systems," *Integrated network management IV* (Sethi, A. S., Raynaud, Y., and Faure-Vincent, F., eds.) (Chapman & Hall, London, 1995), pp. 550–561.
5. Burns, H. S., Chao, C. W., Dollard, P. M., Mallon, R. E., Eslambolchi, H., and Wolfmeyer, P.A. "FAS-TAR™ operatons in the real AT&T transport network," *Conference Record Globecom '93*, pp. 229–233.
6. *TeleStrategies' OSS '98* (Washington, DC, Jan. 12–14, 1998).
7. Gelman, S., and Peck, W. D. "Bringing business information to AT&T network systems through a data warehouse," *AT&T Technical Journal* **75:**2, 68–78 (March/April 1996).
8. Inmon, W. H., Imhoff, C., and Battas, G. *Building the operational data store* (Wiley, New York, 1996).
9. *Journal of Network and Systems Management*, Special Issue: Multimedia Networks/Service Management, **5:**3 (Sept. 1997).

Pillar—Human Factors

Systems are just tools, extensions of human brains and hands. How well the tools work in facilitating comprehension is the result of human factors design efforts. Measuring the ultimate effectiveness of systems—did you save what you needed and can you use it to do work—is the province of human factors engineers.

9.1. MANAGING DATA AND EACH OTHER

The subject of human factors can range across a continuum from the philosophical meaning of the nature of work to the details of optimal screen design. The details have received the most attention, perhaps because of the work of pioneers like Henry Dreyfuss at Bell Laboratories whose elegant design of everyday things enriches our lives, and Steve Jobs and Alan Kay who made computers comprehensible to nonprogrammers.

Human factors specialists do more than design friendly icons, however. They bring two important kinds of knowledge to bear on systems development: first, human abilities and limitations, and second, empirical methods for collecting and interpreting data from people. They define criteria for usability, learnability, and user acceptance in measurable terms.[1] New technology demands much thought about the role of the tool. The distress and aversion many people manifest toward computerization is perfectly rational in the presence of ill-conceived design.

9.1.1. Managing Things, Managing People

It is important to distinguish between the management of things, like data, and people. Say the word *manager* and most people will think of a Peter Drucker clone, that is, an intermediary between the executive and people who actually perform physical work. Drucker's idea was that managers would benevolently pursue the interests of society. Given the turmoil in many large companies and the resonance that Dilbert cartoons have with many people, it would appear that this idea did not work out well, and some major change is happening. Perhaps the groundswell is coming from the as yet imperfectly perceived impression that computers can augment workers, not substitute for them. Knowledge that is freely available flattens office hierarchies and increases the total amount of work it is possible to achieve. A Drucker-type manager need no longer exist when the computer does that job better and faster as, for example, in repair center dispatch.

9.1.2. Knowledge Workers

So what job must be done? It is the job of considered judgment, creative ideas, and efficient problem resolution. Better, faster, more intelligent technology actually serves to minimize the less desirable hostile tendencies in human nature and nurture the more collegial tendencies.[2]

Shoshanna Zuboff foresaw the changing nature of work a decade ago.[3] A powerful new technology fundamentally reorganizes the infrastructure of the work world, especially in the distribution of knowledge in the workplace. We have said several times that the job is to manage data, not databases. Better, faster more intelligent computers and software agents can handle complex problems without frustrating the users. When systems handle the complex as well as the simple parts of their jobs, they nurture the more collegial tendencies in the users. The human factors problem is to design sufficient feedback into broadband networks to allow people to function effectively. When there is a problem in a broadband network, lots of data are generated, describing the problem from every conceivable perspective. The analysis problem is to distill information from the flood of raw data so that decisions can be made. Zuboff remarks,

The psychological discomfort that gives rise to doubt reflects the loss of an immediate knowledge for which there is as yet no replacement. Knowledge had depended in large measure upon the body and its experience of concrete cues. Now the operator feels both personally diminished and weakened by a loss of crucial contextual information. A new sense of certainty depends first upon clarifying the referential function of the data. Deductions are not read off the face of appearances; they are not transparent features of the terminal screen. Rather, they depend upon understanding, and understanding can be developed only from a solid intellective skill base that recognizes what symbols are supposed to represent and that has invented mechanisms for validating that they really do carry such force."[4]

9.1.3. Collaborative Management

As networks become more sophisticated and reliable in taking care of themselves and asking for repair when they need it, it follows that problems, when they occur, will be more complex. Managers who model themselves after Drucker's precepts, however benignly they were intended, stifle the spirit of hypothesis generation, testing, and "playful" manipulation that is above all a collegial one. New network technology can be a catalyst for work that is more flexible and collaborative. This change in the style and substance of work may or may not be desirable for any given individual, but it is important to recognize that more technical leadership in the management of people is part of this package and will cause great differences in the way people collaborate. Nobody wants a computer that will steal his or her job; everybody wants a computer that will enhance personal status. Computerization broadens and expands the available work, but how acceptable this kind of problem solving, decision making work is to people who have no experience or preparation for it will be the staff management challenge.

The overall quality of the body of work will depend on the totality of individual judgments. Military historians understand that the outcome of a war depends on the cumulative effect of very many individual decisions, each made in the heat of battle, in a split second, on the basis of some subset of the whole truth. Decisions that looked good at the time can turn out later to have been not so good. The only tools to minimize the bad collective decision are good data in a form recognizable to many people and distributed widely.

It is a liberating idea that wars, history, and corporate directions are not deterministic, that things do not have to turn out in only one particular way, and choices might be quite different if bits of data were presented more clearly to a wider audience. If information technology improves a work environment by giving a larger and clearer view of the truth of a situation, better individual decisions can be made in moments of crisis. Individual and group judgments can improve. This is a style of management worth cultivating in a rapidly changing world for which there is not much of a template.

One example of people dealing with the unknown and unpredictable is NASA. Human factors specialists on NASA teams preparing for early space missions developed specific techniques of group interaction that fostered optimal group problem solving. It was a collegial style that involved much exchange and repetition of information for maximum clarity. It also involved practice in simulated situations and much hypothesizing of alternatives. The astounding success of the early missions demonstrates the value of human factors applications.

9.2. FACILITATING COMPREHENSION

The remainder of this discussion is devoted to the management of things, specifically networks that shunt around huge amounts of data. As was explained in Chapter 1, network managers handle four networks: one carrying the message traffic itself, a second carrying alarms and measurements from the network elements to the network management system and controls back to the network elements, a third connecting the various network management computer systems together, and a fourth bringing the information to the network managers' terminals and wall displays.

9.2.1. Information Overload

There are places today where network managers' desks are filled with terminals, each accessing a single tool, each having a unique conversation style and idiosyncratic meanings for terms. Bob Lucky of Bell Laboratories amusingly deplores the headaches of "information overload," a phrase that draws thousands of matches on his Web browser.[5] Apparently, many people can relate to the network manager's desk clutter because they themselves feel the burden of fax machines, telephones with call-waiting and call-answering capabilities, car phones, cell phones, PCs, printers, scanners, televisions with both network and cable, radios, and special news services all competing for one's personal attention.

9.2.2. Problems Resulting from Many Systems for One Job

If the individual in ordinary life can feel the burden of too much information burdening the ability to think and make judgments, the network manager's position can certainly be appreciated. It is expensive to buy, learn, and maintain many systems. Many systems means inevitable data overlap, multiple labels for the same data, data incompatibility, and various formats. In crisis, the most needed system is often the least accessible in literal physical placement because the most frequently used system gets the optimum operator position. In crisis, people do what they have already learned best, so less familiar systems confound problem solving.

9.2.3. Data Visualization

Floods of data are useful only when transformed into knowledge, as Bloomberg has done with news or Boeing with its in-house airplane design. Future network management systems will apply expert systems to data that now are largely ignored because of lack of time or analysis tools.[6]

Businesses need a technique for repetitively addressing vast databases which could present information in a form so that nonstatisticians could readily make business decisions, without reducing data to the point where important information is obscured. One technique for extracting information hidden within massive data sets is called *data visualization*. It exploits the sophisticated visual acuity humans have for pattern detection by using color, position, and texture to display encoded data on a graphics workstation.

The process of extracting insights from data involves the following steps:

- Problem definition: delineating the overall framework for the discovery

- Data access: retrieving the data from their warehouse

- Data cleaning: ensuring quality and integrity

- Data mining: extracting information

- Data presentation: showing results from which a decision can be made

- Business impact: specifying how the discovery has resulted in a business decision, modified process, or confirmed result

There are design guidelines for building information-rich visualizations of business data. Visualizations work best if they are task-oriented, domain-specific, and colorful; they must also have high information density controlled by interactive filters, linking to other views, and animation.[7]

Work is also being done to facilitate scientific visualization in a cooperative, distributed setting. Synchronously conferenced collaborative visualization would let multiple users on a network share large data sets and simultaneously view and manipulate data.[8] An ACM Symposium on User Interface Software and Technology was devoted to techniques for data visualization in various disciplines that would make a computer so embedded, so fitting, so natural, that people could use it without even thinking about it.[9]

9.3. REACH OUT AND TOUCH SOMETHING

The Graphical User Interface (GUI) is a particularly apt style of interaction for the tasks performed by network managers because of the possibility of flexible approach to the organization of tasks through multiple windows and graphics.[10] Human capacity to perceive patterns is enormous and GUIs try to capitalize on it by promoting visualization. To understand a problem with visualization, a person must collect data and have some intuition about data interaction.

9.3.1. Government Action

Almost 10 years ago, the government of The Netherlands decided that each Dutch citizen should have the benefit of access to the best that technology had to offer. They realized that massive training programs were impractical. Therefore, computerized tasks had to be as intuitively understandable as turning on a water faucet. Before human factors was developed as a discipline within system design, the only people explicitly trained in the skills of interaction were artists, dancers, actors, and musicians.

Therefore, the government created the School of Interactive Design at the University of Utrecht and staffed it with such people. Professors were forbidden to teach full time because it was critical that they keep current in the business world. The government supported liaisons between businesses and students to accomplish specific projects. In the absence of a body of research into this middle ground, they began to achieve considerable success by using people whose lives were built on the art of communication. They understood that if technology were ever to become more generally useful beyond the domain of professional software designers, software for representing and manipulating complex data visualizations needed to build on an understanding of how people think and work with the world.[11]

9.3.2. Interspace

Interface is an impoverished word, implying far too little of the complex reality of a web of interactions. Terry Winograd of Stanford University coined the word *interspace* [12] to mean the rich interplay of multiple people, workstations, servers, and other devices for which appropriate theories and models must be designed before people can fully use the potential power of networking technologies. The theoretical work in this area will come out of disciplines such as psychology, communications, graphic design, and linguistics, starting from the assumption that a computer system is a shared space for multiple people.

Winograd points out that although there is very abstract theory about workflow and there are very detailed, specialized applications, there is a gap in the middle where no conceptual and computational tools exist to make it easy to bring collaboration into the mainstream of applications. He suggests that the architecture of software, similarly to the architecture of a physical space, shapes the communication patterns of its inhabitants; people are thought to *occupy*, rather than *use*, buildings and, by analogy, inhabit a virtual arena of software.

9.3.3. Integrated Problem Solving

IBM is advancing friendlier systems.[13] The TMN standard is not natural for people; often TMN functions need to be combined and partitioned to solve a problem. Their approach is to stress system usability by deploying standard object-based technology to bind together current products and add new functions in distributed system management encompassing network, configuration, change, performance, and storage.

Hewlett-Packard was one of the first to create a centralized monitoring system to deal with the problem of network crashes. It provided detailed monitoring for the end user, the databases used, the queuing success, every step of the call routing. Network managers needed to resolve problems before they happened, by seeing the symptoms build up. HP built on the signaling network, SS7, because it covered every segment of a network and interfaces to other networks. The product was designed for landline systems, but was being used by wireless carriers by 1993.

Other companies are producing centralized monitoring systems, responding to the demands of surveillance, billing, and customer service. This led to centralizing various networks into large centralized monitoring systems, variously called *network operations centers* or *network management centers*.

GTE created one that consolidates 17 sites in Irving, Texas.[14] AT&T produced a government communications system tying together 1800 state, county, and municipal government locations throughout Wisconsin.[15] New England Telephone in 1990 had a network operations center to provide an overview of all of Massachusetts, Maine, New Hampshire, Vermont, and Rhode Island.[16]

In all of these cases, an old work center strategy of local control centers with independent monitoring systems providing information on equipment and traffic flow was integrated. Graphic display of network surveillance data was critical. UNIX-based workstations provided the speed, graphics, and multitasking capabilities to give a global view of the network. Problem solving required a team approach in these new circumstances.

For example, facilities and switch monitoring were two different jobs. Now both specialists could have easy access to information for crossover in problem solving. Workstation screens were replicated on large wall video displays to allow managers to observe each other's data. Centralization significantly reduced the time it took to locate and define a failure and reduced the analysis because information was simplified, focused, and graphically displayed.

Because there are not enough experts alive to manage complex networks, there need to be products that come with embedded content, expert knowledge, and automated expertise—in other words, expert systems.[17]

9.3.4. Customer Care and Web Agents

Customer care modules integrate the customer contact function for telephony and video services. They provide interfaces for service representatives to access the provisioning and maintenance management modules to use data from the network to determine service readiness. In broadband networks, they also administer video services both at the video provider's server and at the end user's client.

Customer care platforms interface with many systems to reduce the number of screens a service representative must handle. Software encapsulates business process knowledge in executable form. Expert systems capture the decision-making processes concerning customer creditworthiness, combinations of services, and service plans. With object-oriented design and distributed object modules connecting information sources, databases can be updated online, and groups of people can work more easily together.

Growing complexity and the need for mass customization place enormous demands on the architecture of the system. Many customers would also like the ability to provision new services via the Web.[18] The Web provides easy, standard access to multiple servers via the protocol Hypertext Markup Language (HTML).[19] If the customer has a computer, an Internet service provider, a Web browser, and some Internet literacy, he or she can benefit from the Instant Service concept explained in Section 4.2.

Access brings dangers, constraints, and obligations along with convenience. Rather than allow all users access everywhere, *agents*[20] have been developed to handle transactions. An agent is software that is proactive, personalized to the user, and adapted to a specific function.[21] It allows indirect management of the customer's foray into the company's database in that the agent selects and assembles data for the particular customer and oversees any changes the customer makes. Web applications rely on workflow agents, which take a set of tasks and develop an appropriate workflow.[22] Pattie Maes, a professor at the Massachusetts Institute of Technology's Media Laboratory who heads the Software Agents Group, is a powerful innovator in this area. She says that the major challenge is to design the right user/agent interface, combining the issues of understanding and control.[23]

Firewalls[24] must be built at two levels, protocol and application, to deter what are now called *hacktivists*,[25] people who go beyond being mere pranksters to engage in political protest against corporations by attacking their access systems and databases.[26] At the same time, use of the Web increases the need for good data because consistent data must be recoverable over multiple distributed systems. Another consideration is that the Web allows one session per transaction, a constraint that must be accommodated by leaving information in the user's computer to go between transactions.

9.4. SYSTEM EFFECTIVENESS IN HUMAN FACTORS TERMS

Effective systems appropriately assign to the computer those tasks that computers do best and to humans those tasks that humans do best. Determining which is which requires prototyping, user-centered design, user testing, iterative design, usability-based enhancements, and usability testing laboratories, all under the aegis of human factors specialists.

9.4.1. What to Look for in Purchasing Products

System operation should be tolerant of the user's skill and experience. A minimal amount of documentation and training should be required. A system should not be idiosyncratic and should be tolerant of missing or noisy data. The system should have an error strategy that detects errors at the earliest possible point in system processing. The user should not be required to repeat information that the system already has; this is the *minimal input* principle.

These criteria seem self-evident, but they are difficult to satisfy. For example, among widely used software for lay people is Microsoft's Word for Windows, which has been through several iterations, yet its use still requires a knowledge of programming folklore and the purchase of a 2-inch-thick manual. Unless human factors specialists are involved in the earliest stages of design specification, the difficulty of the art of programming causes these major criteria to be obscured. Table 9.1 summarizes the usability principles to which a buyer should demand any software or system product adhere.[27]

9.4.2. Simple Guidelines for Managing Development

If the system or software will be developed, not purchased, the opportunity exists to do it correctly in human factors terms from the start. The following are some simple guidelines a manager can consider to allow the process to occur, but the primary, overarching principle is that human factors work requires specialists. Design does not happen on the basis of programmers' intuition, managers' instincts, or luck. Systems that are difficult to use are an ongoing expensive exercise in programming arrogance and are unworthy of computer design professionals.

1. Hire human factors professionals. There is a large body of research in human perception, measurement, and user-centered design that can be practically applied.

2. Integrate the human factors specialist into the design team. If there is a human factors department in your organization, view it as a resource to support the specialist devoted to your design team, not as a marketing frill to mention in a sales brochure.

Table 9.1. System/Software Usability Principles

Principle	Explanation
Speak the users' language.	Use words, data labels, and concepts familiar to the user. Present information in a natural and logical order in the user's context.
Be consistent.	Indicate similar concepts through identical terminology and graphics. Adhere to uniform conventions for layout, formatting, typefaces, and labels.
Minimize the users' memory load.	Rely on recognition, not recall. Do not force user to remember information across documents. Rely on human factors specialists to determine appropriate levels of memory demand based on context and user skill.
Design for flexibility and efficiency.	Accommodate a range of user sophistication and diverse user goals.
Design aesthetic and minimalist graphics.	Create visually pleasing, efficient displays that capitalize on known human capabilities in pattern recognition, color differentiation, and so on.
Recognize the power of *chunking*.	The capacity of human short-term memory is small, but can be amplified by grouping subsets of information around keywords, by completing single thoughts in one document, and by keeping tasks short but information-rich.
Structure progressive levels of detail.	Organize information hierarchically, with more general information appearing before more specific detail. Allow the user to stop when sufficient information is received.
Facilitate navigation through program structure.	Allow user to determine current position in the program structure. Make it easy to jump among related tasks. Make it easy to return to an initial state.

3. Expect the human factors specialist to test the product on specific, measurable criteria for usability and to develop testing scenarios. Such criteria include rate of human error, time to learn specific functions, speed of task performance, subjective user satisfaction, and human retention of functions over time. Expect to devote resources to prototype trials, field trials, and user surveys.

4. Explicitly assign resources to address usability. Laboratory test time and space must be allotted for evaluating prototypes and system releases on the basis of human factors criteria.

5. Do not relegate the human factors specialist to the publications department. Though good graphic design of supporting documentation is a part of system usability, the major contribution of the human factors specialist is identifying software usability problems and working with programmers to arrive at functional solutions.

6. Support technology transfer for human factors specialists as you would for any other member of the design and development teams.

9.5. HOW MUCH SHOULD THE SYSTEM DO?

One of the most difficult design tasks is to anticipate the ultimate uses of a system. It appears that users will happily manipulate a simple system to provide outcomes that were not conceived by its designers (in the law, this is a loophole), but they expect that if it can do some of the tough analysis, it should be able to do it all. Zuboff identified this as a "trough of disillusionment",[28] which hints at a certain anthropomorphizing of the machine. Users want it to communicate and "think" like a human being.

9.5.1. Work Still Needs to Be Done

The U.S. government, like The Netherlands government, has also encouraged research concerning the accessibility of knowledge. Many scientists have come to understand the need expressed in the Zuboff quotation at the beginning of this chapter callmg for clarifying the referential function of the data.

Table 9.2 summarizes a compilation by Mark T. Maybury of Mitre Corpora-

Table 9.2. Areas for Continuing Research in Intelligent Interfaces

Area	Long-term research
Text processing	Scalable, trainable, portable algorithms; document-length text generation
Speech processing	Large-vocabulary, speaker-independent systems for speech-enabled interfaces
Graphics processing	Automated, model-based creation and tailoring of graphical user interfaces
Image/video processing	Visual information indexing and extraction (e.g., human behavior from video)
Gesture processing	Intentional understanding of gesture; cross-media correlation (with text and speech processing); facial and body gesture recognition
Multimedia integration	Multimedia and multimodal analysis; multimedia and multimodal generation; investigation of less-examined senses (e.g., taction, olfaction)
Discourse modeling	Multimedia and multimodal analysis
User modeling	Hybrid stereotypical/personalized and symbolic/statistical user models
Visualization	Multidimensional visualization; multimedia (e.g., text, audio, video) visualization
Collaboration tools	Experiments to predict impact of collaborative technology on current work processes; tools to analyze video session recordings; flexible, workflow automation
Intelligent agents	Shared ontologies; agent integration architectures and/or control languages; agent negotiation

tion of recommendations for research in intelligent interfaces; clearly much remains to be done in many areas. GUIs and wall displays are the mere beginnings of capitalizing on the multiple senses and intellectual abilities of humans, amplified by technological power. The better the processing of information and the smarter the machine is about assessing the user, the more efficient becomes the support of frequently changing tasks.[29]

9.5.2. Conclusion

It is appropriate that systems support human collaboration. The field of computer-supported cooperative work explores the nature of joint work through technologies that include shared editors, group discussion support tools, and awareness systems. The change from stand-alone to networked computers is transforming computers from data boxes into communication devices.[30]

REFERENCES

1. Designing the Human Interface Issue, *AT&T Technical Journal* **68:**5 (Sept./Oct. 1989).
2. Pascale, R. T. "Zen and the art of management," *Harvard Business Review* March–April 1978, pp. 153–162.
3. Zuboff, S. *In the age of the smart machine* (Basic Books, New York, 1988).
4. Zuboff, pp. 83–84.
5. Lucky, R. W. "Just right," *IEEE Spectrum* Jan. 1997, p. 21.
6. Bernstein, L., and Yuhas, C. M. "Expert systems in network management—The second revolution," *IEEE Journal on Selected Areas in Communications* **6:**5 (June 1988).
   Bramer, M. *Practical experience in building expert systems* (Wiley, New York, 1990).
7. Eick, S. G., and Fyock, D. E. "Visualizing corporate data," *AT&T Technical Journal* **75:**1, 74–86 (Jan./Feb. 1996).
8. Anupam, V., Bajaj, C., Schikore, D., and Schikore, M. "Distributed and collaborative visualization," *Computer* **27:**7, 37–43 (July 1994).
9. *Seventh Annual Symposium on User Interface Software and Technology '94* (ACM Press, Nov. 2–4, 1994).
10. Cunningham, J. P., Blewett, C. D., and Anderson, J. S. "Graphical interfaces for network operations and management," User Interface Design and Development Issue, *AT&T Technical Journal* **72:**3, 57–66 (May/June 1993).
11. Shedroff, N. "Interfaces for understanding," *More than screen deep* (National Academy Press, Washington, DC, 1997), pp. 252–259.
12. Winograd, T. "Interspace and an every-citizen interface to the national information infrastructure," *More than screen deep* (National Academy Press, Washington, DC, 1997), pp. 260–264.
13. Vaughan, N. "Networks and systems still seeking suitable frameworks," *Software Magazine* Oct. 1994, pp. 53–58.
14. Blake, P. "Systems you can't knock," *Telephony* Feb. 23, 1998, pp.46–50.
15. Cosgrove, J. G. "Wisconsin brings network management under control," *Telephony* Nov. 10, 1986.
16. Andrianopoulos, C. "Welcome to the telecom future," *Telephony* Dec. 3, 1990.
17. Herman. "Let's focus on content, not GUI's," *Business Communications Review* Aug. 1995, p. 52.
18. Myers, J., and VanderWall, J. "Lipstick on a bulldog," *Telephony* Oct. 27, 1997, pp. 34–40.
19. Lee, B. H. "Embedded Internet systems: Poised for takeoff," *IEEE Internet Computing* May/June 1998, pp. 24–29.
20. Krulwick, B. "Automating the Internet: Agents as user surrogates," *IEEE Internet Computing* July/Aug. 1997, pp. 34–38.

21. "Pattie Maes on software agents: Humanizing the global computer," *IEEE Internet Computing* July/Aug. 1997, pp. 10–19.
22. Huhns, M. N., and Singh, M. P. "Workflow agents," *IEEE Internet Computing* July/Aug. 1998, pp. 994–996.
23. Schneiderman, B., and Maes, P. "Direct manipulation vs interface agents," *Interactions* Nov.Dec. 1997, pp. 42–61.
24. Amoroso, E., and Sharp, R. *PCWEEK Intranet and Internet firewall strategies* (Ziff-Davis Press, Macmillan Computer Publishing, Emeryville, CA, 1996).
25. Harmon, A. "'Hacktivists' of all persuasions take their struggle to the Web," *The New York Times* Oct. 31, 1998, pp. A1, A6.
26. Cheswick, W. R., and Bellovin, S. M. *Firewalls and Internet security: Repelling the wily hacker* (Addison–Wesley, Reading, MA, 1994).
27. Levi, M. D., and Conrad, F. G. "A heuristic evaluation of a World Wide Web prototype," *Interactions: New Visions of Human–Computer Interaction* **3:**4, 50–61 (July/Aug. 1996).
28. Zuboff.
29. Maybury, M. T. "Intelligent multimedia interfaces for 'each' citizen," *More than screen deep* (National Academy Press, Washington, DC, 1997), pp.246–251.
30. Terveen, L. "Computer-mediated collaboration," *More than screen deep* (National Academy Press, Washington, DC, 1997), pp. 307–314.

10

Economic Impact and Cost Studies

Now that we have reached the final chapter, let's review the questions that lead to developing the value proposition for any enterprise. Everyone who wants to develop a business case has to satisfy certain conditions concerning their new product or service. The customer for the product has to be identified and profiled. The technology supporting the system or product or service has to be stable and be tested for robustness under a real work load. The customer has to understand how it fits into current operations. The product or service has to meet human factors standards of ease of use, documentation, and training in order to be used as designed. Finally, the customer needs to see the economic benefit in terms of payback. Previous chapters have explored the first four questions. The fifth question, economic benefit, is the subject of this chapter. We will describe a method for showing economic benefit in controlled terms so that various options for installing a new product or service can be explored and weighed by the customer.

10.1. WHAT CONSTITUTES ECONOMIC BENEFIT?

A good business case shows a combination of new revenues and significant operations savings. The telephone industry has not historically fallen victim to the fallacy of thinking that capital reductions can improve productivity. In fact, productivity was accomplished by substituting capital for labor, as, for example, in Direct Distance Dialing (DDD) and credit card calling. In the 1970s, OSSs were introduced to ease and speed the work of employees. These OSSs had databases to keep track of all customer accounts, customer requests for service, wires, cables, repeaters, plug-in circuits, trucks, addresses, telephone numbers, bills, and payments. They are used in planning, designing, and administering the network. With improvements in technology, equipment shrinks in size and price while it grows in power. For example, one-sixth as many people handled six times as many calls if one compares 1985 to 1950. This impressive productivity leap was accomplished by substituting capital improvements for labor.[1] Some of the new revenues from broadband services are apparent and many are as yet undeveloped, but the topic of all of the possibilities for new revenues is very broad. What follows is a discussion of the second key parameter for showing economic benefit, operations savings.[2]

10.2. ANALYZING OPERATIONS SAVINGS

The telephone industry has always made steady gains in productivity and profitability, accompanied by steady declines in the cost of calls because it has always used new technology to reduce operating cost, progressing from manual work, to flow-through to dynamic operations. Between 1950 and 1990, the number of access lines increased fourfold while the number of employees needed to install, operate, and maintain them grew by less than 40%. The industry lends itself to computerization and mechanization because the uniformity of procedures allows careful, objective, quantitative monitoring of efficiency and because thorough testing with real work loads before cutting over new systems is standard practice.[3]

The specific dollar amounts used throughout this chapter come from the work of Berkowitz, Snow, and personal experience. The floor debate during the February 1996 World Loop Conference confirmed their general applicability. A generic telephone company expense model (see Fig. 10.1) would show total expenses divided roughly equally among corporate overhead, depreciation and amortization, product and services support, and customer operations and sales. Those total expenses needed to operate the telephone company would cost $450 per line per year. Within that $450 and coming from both the product/services support and customer operations/sales area, the cost to operate a line is $230. The $230 in operations expenses is spent on repair and maintenance, switching and network operations, customer service, engineering, data administration, account management, and provisioning and installation.

The goal is to reduce operations expenses through the introduction of modern network elements and the use of embedded network management. Much of the cost structure has built up over time to take care of the evolution of the network equipment. Because network equipment typically had a 20-year life, telephone company craft had to keep up with old technology while learning the new. This led to an inefficient business structure with many redundant activities.

The telephone companies divide the outside plant network operations into work centers and systems dealing with POTS and those dealing with special services. The annual cost to provision a POTS line is $35, which is spent on outside plant cross-connects and rearrangements, customer contact, order creation, and account management, MDF connections, and assignment, order and dispatch administration as shown in Fig. 10.2. Special service provisioning for anything other than POTS is more expensive and ranges from $125 to $250 depending on the service being installed. When multiple visits to the customer and plant rearrangements are required to set up the service, these costs may be even higher. Maintenance for a POTS line adds an additional $30 per year in outside plant repair and rehabilitation, customer contact and diagnosis, MDF connections, and testing, analysis, diagnosis, and dispatch administration as shown in Fig. 10.3. Special service maintenance may double or triple that basic cost depending on the preventive maintenance phi-

Total expenses: $450 per line

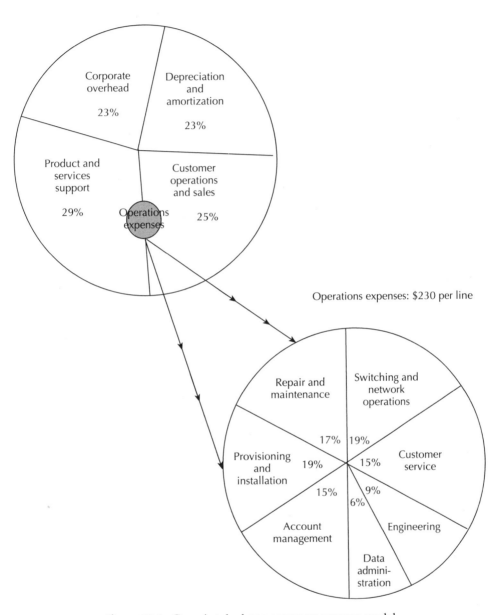

Figure 10.1. Generic telephone company expense model.

losophy of the company and the investment in special test points and test equipment. Table 10.1[4] shows how the $35 per line to provision POTS is calculated. For an economic study, similar analysis to that done to achieve these POTS figures would be necessary for each type of service installed on specific outside plant technologies and equipment configurations.

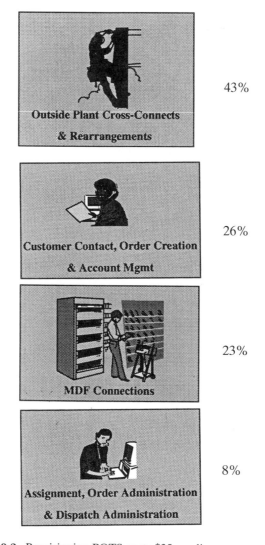

43%

26%

23%

8%

Figure 10.2. Provisioning POTS costs $35 per line per year.

10.3. STRATEGIES FOR OPERATIONS SAVINGS

First, a distinction should be made between savings and benefits. Savings (*S*) are the difference between the dollars needed to run the present method of doing business and the similar dollars needed to run the new method. Benefits (*B*) are the difference between savings and the dollars needed to acquire or develop, then install the new method into the environment. Let *P* stand for the dollars needed to run

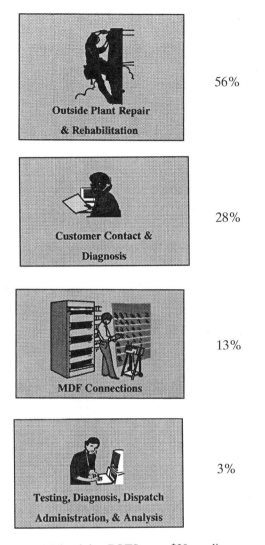

Figure 10.3. Maintaining POTS costs $30 per line per year.

the *present* method of doing business and N stand for similar dollars to run the *new* method. Let I stand for the dollars to acquire or develop, then *install* the new method into the environment. Then, $S = P - N$ and $B = S - I$. Another way to think of benefit is that $B = P - (I + N)$.

Telephone companies can follow several strategies to reduce the cost of providing service and derive operations savings (see Fig. 10.4).

Table 10.1. Cost Analysis for Provisioning POTS (Copper)

	Occur factor (%)	Task time (hours)	Labor rate ($/hour)	Subtotal ($)
Order administration				
Customer Contact	100.0	0.14	35.00	4.87
Initiate SO	100.0	0.10	35.00	3.50
Assign TN	12.0	0.10	35	0.42
Order tracking	100.0	0.15	35.00	5.39
Assignment				
Facilities	4.1	0.20	40.00	0.32
Equipment	4.1	0.20	40.00	0.32
Plug-in procurement				
Travel	0.0	0.50	60.00	0.00
Obtain plug-ins (OP)	0.0	0.10	60.00	0.00
CO provisioning control				
Administer work force	10.8	0.10	50.00	0.54
Allocate work items	10.8	0.10	50.00	0.54
Dispatch	5.4	0.10	50.00	0.27
Memory administration				
Generate recent change	15.0	0.10	40.00	0.60
CO craft				
Travel	5.4	0.50	60.00	1.62
Obtain plug-ins	0.0	0.10	60.00	0.00
Connect/install	10.8	0.20	60.00	1.30
Test	10.8	0.10	60.00	0.65
OSP provisioning control				
Administer work force	8.1	0.10	50.00	0.41
Allocate work items	8.1	0.10	50.00	0.41
Dispatch	4.1	0.10	50.00	0.20
OSP craft				
Travel	8.1	0.50	60.00	2.43
Connect/install	8.1	1.00	60.00	4.86
Test	8.1	0.20	60.00	0.97
Total				29.61
Annual order rate/line	1.2			
Annual cost/line				**35.53**

10.3.1. Integrate Access to OSS Functions and Data

For copper networks, telephone companies integrate access to all of their operations systems and to their various databases. This is done with clients who have access to a number of legacy systems and can translate the screen formats and data descriptions for the use of the employees. Providing this type of integration to their

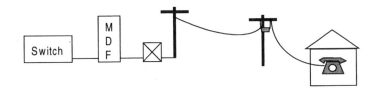

Strategies for Copper Networks	Barriers to Implementing Strategies

Customer Services
  Integrate access to OSS
    functions and data
  Soft Dial Tone processes
Provisioning / Installation
  DIP/DOP
Preemptive Maintenance
  Fix it before the customer
    finds it.
	Data do not match network
	  Latency of updates
	  Pressures preventing craft diligence
	  Data update system unavailable
	    when needed
	Difficult to achieve high DIP/DOP levels
	  Feeder capacity churn
	  Fill rates approach 95%
	  Poor data integrity causes reworking
	Impracticality of preemptive maintenance
	  Bottom line savings must be inferred

Figure 10.4. Methods for deriving operations savings in copper networks: strategies versus barriers to implementation.

own service representatives has been successful in eliminating entire work centers of service order typists.

10.3.2. Instant Service

In the 1990s, some telephone companies introduced the concept of Instant Service, which allowed the service representative to activate or modify service during the customer contact. Automatic flow-through assignment was extended to include automatic recent change in the switches so when there were facilities in place and the customer had good credit, the phone was turned on. This capability was extended again with a concept called Soft Dial Tone to let customers automatically activate their own service. When a customer moves out of a residence, the telephone company leaves all of the equipment used by that customer in place in anticipation of the next customer requesting similar service. With Soft Dial Tone, a new customer moves in, plugs in a phone, and gets a special dial tone that allows calls only for emergency services and the business office. A call to the business office can activate service through the Instant Service feature with no human contact whatsoever. Soft Dial Tone with Instant Service eliminates all telephone company employee intervention from the process of establishing service. This does cut out a customer contact point where additional services can be explained and offered for sale, but may prove useful under certain circumstances such as Web access.

10.3.3. Provisioning/Installation

To make Instant Service possible and to reduce the need to dispatch technicians to the field, telephone companies have adopted the policy of leaving established equipment and connections in place. When the central office equipment is left in place, the policy is called Dedicated Inside Plant (DIP). When the equipment and services are left in place outside the central office, the policy is called Dedicated Outside Plant (DOP). When both policies are in place, the term is *home run*. This policy has several goals; in order of priority for creating savings, they are as follows:

- Reduce the need to dispatch expensive craft people in trucks to the field, but still provide timely service

- Reduce the investment in installation equipment

- Reduce the opportunity for error to be introduced to the physical network and the database

10.3.4. Preemptive Maintenance

The concept of preemptive maintenance suggests that it is cheaper to find and fix problems before the customer experiences trouble. The corollary is that a lower fill rate reduces churn and makes problems easier to find and resolve. Buried in the complicated problem of managing plant equipment is a perfect example of how misguided attempts to minimize capital investment can cause ballooning costs. When plant utilization is pushed toward 95% in the mistaken effort to squeeze every bit of use from a network, there is hardly ever a spare pair where you need one to install new service. If dedicated plant policies are not enforced and there is a spike in demand for new services, craft people must rearrange plant. When there are floods, windstorms, or equipment failures, more rearrangement must be done. The result is tremendous churn in installation.

One new line can cause three plant changes. Fallible human beings tend to introduce one error into the plant or the database for every three rearrangements. Therefore, in the worst case, one new line = one error. A churn rate of 5% leads to a 25% worst case (20% best case) increase in provisioning and maintenance costs (see Fig. 10.5). The devastation to the database is comparable.

Rearrangements are always crisis-driven, but it is neither possible nor desirable to freeze a dynamic system in order to maintain good data and neat hookups. The solution is new and better human factored equipment to reduce errors and a fill rate of less than 85%. Studies have shown that customers are willing to pay 10 to 20% more for reliability. This speaks strongly to the case for broadband with its packet switching to the home and trunklike signaling that allows dynamic, consistently available provisioning. One might tie the expense savings, expense avoidance, new revenues, and keeping a satisfied customer base to offsetting the expense of broadband.[5]

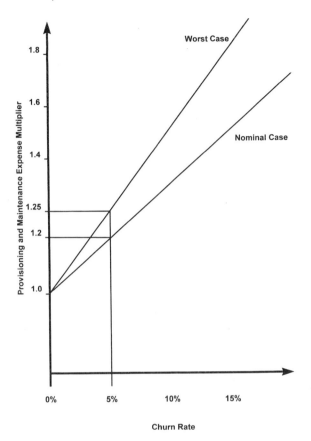

Figure 10.5. Cost of churn in outside plant.

10.4. WHAT'S EASY, WHAT'S HARD

Predicting software costs, schedules, and performance remains a serious problem, but the successes have far outweighed the failures. One system that manages the inventory of the expensive plug-in circuit cards used in all of the modern digital telephone equipment was so profitable that it alone paid for the development of all other OSSs. This system, which is still in use today, managed to find so many "lost" plug-in cards that the telephone companies were able to grow their customer base by 3% without buying any new equipment.

On the other hand, outside plant automation has been difficult to achieve. It often takes twice as long to update the records showing where changes in the plant were made than to do the actual work. This update time detracts from providing customer service and leads to neglecting the update task. We have the image (thanks to a wonderful ad campaign in the 1970s) of a noble lineman hanging off a telephone pole in the pouring rain, waiting for someone from the assignment bureau to issue a new pair. In real life, this hapless lineman picks any available pair and makes a mental note to call in the change tomorrow. When he shows up for work the next day, there is another pressing problem and the database is the orphan child again. Estimates are that there is a 5% error rate in the central office records, a 10% error rate in the outside plant records, and as much as a 25% error rate in the maintenance records.

Even with the best of intentions, it is difficult to leave telephone equipment dedicated to a specific location. The temptation to use it for an actual new customer when it is sitting there idle waiting for a prospective customer is too great to resist. From long practical experience, telephony engineers have learned to plan for an 85% fill rate for the feeder plant. If they try to optimize further, they compromise the ability of the craft to adhere to dedicated inside and outside plant policies and still meet customer requested dates for service.

However difficult it may be to achieve, reducing trouble reports by preemptive maintenance ultimately reduces costs. Remote automated testing for switched data and ISDN can save $30 per line per year. Advanced remote testing for DLS-based lines saves $9 per line per year. Remote customer premises fault isolation on all switched access lines saves $3 per line per year. Network architectures that slow churn have inherent operations advantages that can be exploited by sophisticated OSSs. Products with integrated functions can replace a local service provider's existing copper with either a hybrid fiber/coax or a totally fiber network. A local service provider's collection of operations systems can be scrapped in favor of a single design using network elements to support a concept called *dynamic network operations*. The solution often consists of multiplexing subscriber loop carrier systems in the central office, fiber to the neighborhood, coax in a buslike structure to the home, and attaching an NIU outside the home. The NIU splits the signals coming over the coax into POTS and video for inside the home. Figure 10.6 shows savings based on weighted averages for a typical line that result from dynamic network operations.

The intelligence in the network elements, including the NIUs, allows for many operations improvements, among them being Instant Service provisioning, preemptive maintenance, and geographically directed dispatch. A model of a telephone company with 70% residential and 30% business customers operating with the new architecture reveals annual savings of $65 per line. The saving is itemized as follows: installation $30, maintenance $22.50, assignment $5, engineering $5, and OSS administration $2.50. A broadband computer support system can replace legacy maintenance and provisioning operations systems. The problem here is the cost of deploying the network technology and getting the permits to allow deployment of the technology. Cable companies have the advantage of in-place cable networks. They do not have to get permits or rights of way. Their problem is to upgrade their cable plant so that it can reliably provide telephone service. Most cable networks were originally designed for one-way transmission and reliability was very poor. With the introduction of telephony using hybrid fiber coax technology, the cable companies could leapfrog the local telephone companies. Telephony will introduce intelligence to the cable network with self-identification, self-test, and automatic call set-up features that will keep operations costs in line.

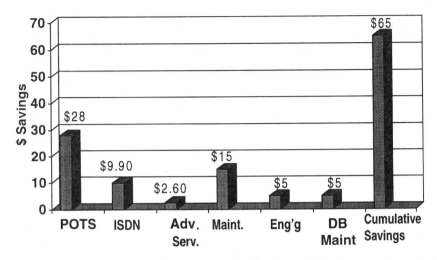

Figure 10.6. Dynamic network operations savings based on weighted average for typical line.

10.5. WHY BROADBAND?

Figure 10.7 shows that the strategies to reduce costs for broadband networks, which correspond to the copper network strategies in Fig. 10.4, focus on customer services, provisioning/installation, preemptive maintenance and engineering. The technology that enables the efficiencies in each is that software replaces MDF, there can be remote control of active elements, an intelligent terminal is near the customer site, and remote service activation is possible. Overall, engineering is made simpler.

The Host Digital Terminal (HDT) interfaces to a fiber node which then translates the signals from the photons on the fiber to electrons on the coaxial cable. This terminates in special NIUs containing data that can be used for automatic service activation, testing, and configuration management. These networks offer the great advantage of being able to integrate network management and other OSS functions into the network itself. The data will be synchronized automatically to the actual condition of the network and Soft Dial Tone can be built into the network.

The use of the TSI capability of newer network protocols and switch interfaces eliminates the need for the costly administration of the traditional MDFs. The network automatically assigns a slot to the originating equipment in the switch. This provides for electronic coupling of the outside plant to the inside plant. Figure 10.8 shows a breakdown of where the $65 per line savings are obtained.

The vital first step in containing operations costs is to make sure that there is enough network capacity to handle traffic growth, port growth, and to rearrange routes in the network. Broadband networks are amenable to capacity engineering and traffic concentration techniques. The design focuses on service type and builds in capacity for operations data, relieving the engineer of a single-minded devotion to network placement. Algorithms are built into the OSSs to detect congestion and manage it. With the assignments tied to bandwidth use and not solely to fixing

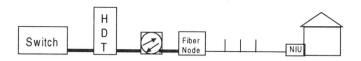

Strategies for Broadband Networks	Technology Enablers
Customer Services 　Integrated OSS 　Data synchronized to network 　Soft Dial Tone Provisioning / Installation 　Dynamic provisioning 　Bandwidth on demand 　High bandwidth close to customer 　　reduces churn Maintenance 　End to end surveillance 　Preemptive maintenance 　Automatic diagnosis Engineering 　Automatic rearrangements 　Automatic loop makeup	Software to replace MDF Remote control of active 　elements Intelligent terminal near 　living unit Remote service activation

Figure 10.7. Methods for deriving operations savings in broadband networks: strategies and technology enablers.

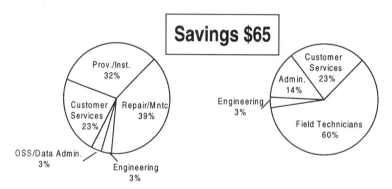

Savings by Function	Savings by Craft
Customer Services 　Lower trouble rates 　Higher productivity Provisioning / Installation 　Low dispatch rate 　No MDF work 　Lower administrative overhead Repair / Maintenance 　"Hands off" means less trouble 　Better diagnosis means faster repair	Engineering 　Fewer activities 　Less costly construction OSS / Data Administration 　Synchronization with network 　results in fewer errors 　TSI eliminates MDF work

Figure 10.8. Origins of savings realized through use of TSI capability.

90% fewer customer trouble reports
95% shorter provisioning interval
95% shorter restoration interval
50 to 65% fewer employees

Tomorrow's Simplicity

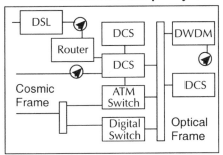

Digital switch saves $5/L/Yr
Digital cross-connect switch (DCS)
 saves $10/L/Yr
Digital subscriber line (DSL)
 saves $4/L/Yr

Figure 10.9. Network modernization produces cost reductions.

resources to each customer, the outside plant engineer can design a tuned network for the expected service load. This approach saves capital investment and eliminates much of the churn in the outside plant network.

10.6. HOW TO GET THERE FROM HERE

That network modernization is an ongoing mandate for telephone companies is a given. The technology exists to simplify older technologies in 5ESS-2000, DACS IV-2000, and SLC-2000. Figure 10.9 shows how a complex network could be simplified by using these tools. What remains is the question of how to get from here to there with the most efficiency and least pain. D. F. Snow has done much work at Lucent Technologies to develop a model for access modernization to analyze alternatives for this purpose.[6] The model includes both noneconomic factors and economic viability factors. The noneconomic factors are governance, which includes all of the elements of government oversight of physical public networks such as spectrum and orbit allocations, rights of way, franchises, and so on, and strategic fit, which includes the relative strengths and weaknesses of a project, the opportunities in the business environment, and the project's priorities for addressing these. The economic viability factors are cash flow and key drivers as defined by the financial community. What follows is a brief discussion of the economic viability factors considered in the model.

Cash flow is money generated over time. For service businesses, the final value at the end of the time period of the study is subjective and depends on the business's competitive position. Therefore, the model assumes a growth or operating strategic view, in other words, a perpetuity model. Free cash flow is used in this analysis:

Free cash flow = sales − (cash operating expenses + taxes without interest income
+ change in customer deposits + change in working capital
+ capital expenditures)

The fundamental formula of the model is as follows:

Project value = study period value + terminal or residual value

10.6.1. Definition of Formula Terms

Study Period Value

The present worth of a cash flow over a period of time is a straightforward "time-value-of-money" calculation. Given a discount rate, the present worth of the cash flow in year x is calculated:

$$PV_x = \text{cash flow}_x/(1 + \text{discount rate})^x$$

where PV_x is the current or present value for the cash flow in year x, cash flow$_x$ is the cash flow in year x, and discount rate is the composite of equity and debt rate.

Terminal Value

Terminal value is calculated under an assumed operating strategic outlook at the end of the study interval. The method used is referred to as a perpetuity method. Here the business is assumed to be an ongoing concern, capable of generating returns from a given cash flow. Furthermore, it is assumed that after the end of the study period, the business will *earn*, on average, a rate of return equal to the cost of capital on new investments. The value at the end of the study period is then

Terminal or residual value = (perpetuity cash flow)/(cost of capital)

Present value must be discounted as per the previous PV formula.

Discount Rate

An appropriate discount rate must be applied to arrive at a discounted cash flow. For this evaluation, the rate is the weighted average of the cost of debt and equity capital. The cost of debt is straightforward. The cost of equity is more difficult to estimate. In essence, it represents the implicit rate of return required to attract investors to purchase a firm's stock as well as to induce existing shareholders to hold their shares. Conceptually this may be modeled as follows:

$$\text{Cost of equity} = \text{risk-free rate} + \text{equity risk premium}$$

where

$$\text{Risk-free rate} = \text{real interest rate} + \text{expected inflation rate}$$

$$\text{Equity risk premium} = \text{expected return on market} - \text{risk-free rate}$$

10.6.2. Applying the Model to Three Alternatives

The model was employed to step a generic local exchange carrier through the following three alternatives for upgrading its plant to lower costs, raise revenues, and offer broadband services:

- Flash cut the old technology with the embedded base over the first year. Borrow heavily to pay for it. Aggressively pursue new revenue opportunities starting in the second year.

- Gradually roll-over the embedded base of old technology with new technology and initiate new revenue opportunities as soon as the rollover is complete. All new growth is on a new technology platform, deploying hybrid fiber coax cable to replace drop wire and deploying fiber to the serving terminal.

- Keep the old technology for the embedded base while placing all growth on a new technology that offers operational improvements and the capability for new revenues sometime in the future.

10.6.3. Combinations of Six Key Financial Drivers

Six key financial drivers are also examined in various combinations of relationships. They are defined as follows:

- *Footprint*: 100,000 households growing at 2% per year

- *Service penetration*: assume all alternatives identical

- *Market share*: show for each service capability against each modernization alternative. Consider implications resulting from service bundling and market share advantages from early market presence in the broadband/ multimedia area.

- *Margins*: gross margins are derived from the difference between revenues and operating expenses.

- *Revenue per subscriber*: normalize revenues on a per unit basis, e.g., base year 1994 with values of $600 per line for telephony, $350 per customer for CATV, and $2000 per customer for broadband/multimedia.

- *Capital*: level of capital expenditures on a per line or customer basis from loop only perspective (sufficient for a differential view but underestimates the capital requirements on a stand-alone basis).

10.6.4. Results

The flash cut alternative shows as being the best bet financially, but it is surely the most painful in practical terms. Another study finds that for all examined alternatives, the cost of a broadband upgrade is greater than or equal to establishing something like the existing network from scratch with payback periods of at least 5 to 10 years:

Results illustrate that the average capacity demand in the access network is a very important parameter, since the actual capacity provided is a differentiator between the different technology options and their associated overall investments. Copper/DSL vs coax/cable modem technologies are comparable for "take rates" up to 30%; higher penetrations: cable modem technology seems to have cost advantage but a limitation in traffic capacity compared to DSL technologies. Optical fiber alternatives need "take rates" greater than 50% and more than 50 subscribers per optical node to be economically justifiable. Fiber rollout seems to be a key strategic decision since the costs of extensive fiber deployment are strongly related to civil works costs [digging, ducting, surface repair] and cost reduction relies heavily on optical network termination customer sharing.[7]

The basis for improved economics is a 50% reduction in operating expenses, which is possible if network management is moved into the network itself. If these savings cannot be obtained, postponing fiber in the loop will have cost advantage over hybrid fiber coax. Under this condition, cable modems in CATV present the best cost base for service beyond POTS.

10.7. CONCLUSIONS

The future belongs to broadband network service providers. They will be able to offer faster networks, more services, more reliable networks, and lower prices than telephone companies, with predominately copper plant. Cost savings using dynamic provisioning, reducing the number of components, and anticipating maintenance can be as much as $100 per line per year relative to present methods of operation in the copper plant. These savings can be used to pay for the installation of broadband loops.

Companies with large investments in copper plant can evolve to a broadband plant in steps. First, they could replace their feeder plant and main distributing frames with broadband network elements. Then they could migrate broadband distributing networks close to the customer. At this writing, fiber broadband networks are economically feasible when they serve 200 to 500 residential customers. From the point of the fiber concentration, coaxial cable could be used to reach the customer and provision broadband service. If copper plant is used from the end of the fiber, broadband service will not be available until the fiber can be extended to serve 1 to 10 residences. This strategy is called making the plant "broadband ready." It does not offer the immediate economic advantage of a broadband network, but it positions the telephone company to eventually provide the service.

A mix of broadband, broadband ready, and midband services can be deployed. Trying to have a single strategy delays decision making as no one approach fits the complexity of telephone company markets. Cable companies can provide broadband telephone service by adding multiplexed services on their coaxial cable. Not only could they add services, but they could use the inherent network management features of a broadband network to improve the reliability of their other cable services.

There are indications that if telephone companies and cable companies are too slow to embrace broadband, government will step in to fill the void because the public requires this service. The city governments of Glasgow, Kentucky, in 1989 and Tacoma, Washington, in 1997 installed fiber-optic cable in order to democratize the Internet and offer competition to monopolistic cable companies. In Palo Alto, California, the city government invested a mere $2M to build a 15-mile-long fiber optic ring in partnership with Digital Equipment Corporation which for its part created a new Internet Exchange, the world's biggest on-ramp to the information highway. Every home and business is within 1 mile of this ring. Digital avoided the issue of rights of way to lay cable by striking an agreement with the city to lay fiber through city conduits. This synergy of culture, business, and technology is attracting industries that demand large amounts of Internet bandwidth; the Digital hub can be expanded to support the equivalent of 87 million simultaneous telephone calls. There are similar projects under way in California for Los Angeles, San Diego, and Anaheim; the project is already completed in Winston-Salem, North Carolina.[8]

In private enterprise, the biggest player is Qwest Communications International, Inc., founded in 1988 and trading publicly since June 1997. It is engaged in building an advanced fiber-optic network reaching more than 18,000 miles, coast to coast in the United States. Qwest anticipates offering its own services and selling capacity to other telecommunications companies.

The discussion and turmoil surrounding the complexity of introducing broadband services are part of the continuing development of the communications industry and it is consistent with its past. If it seems more difficult today, it is simply that the future is happening faster than it used to.

REFERENCES

1. Nousaine, T., and Brant, S. "A case for capital investment," *Telephony* Feb. 2, 1987, pp. 58–60.
2. Elmquist, M. D. "Operations support for full-service networks," *Multimedia over the broadband network: business opportunities and technology* (International Engineering Consortium, 1996), pp. 145–149.
3. Landauer, T. K. *The trouble with computers: usefulness, usability and productivity* (MIT Press, Cambridge, MA, 1996), pp. 64–72.

4. Snow, D. F. Lucent Technologies Inc. Unpublished. Used with permission.
5. Bernstein, L., and Yuhas, C. M. "The case of the creeping error, or 1:3:3:1," *Journal of Network and Systems Management* **5**:2, 105–107 (June 1997).
6. Snow, D. F. "A financial framework for evaluating major network investments: A framework for evaluating provider investments including examples of LEC access modernization alternatives," Lucent Technologies Inc., unpublished, July 14, 1997. Used with permission.
7. Ims, L. A., Myhre, D., and Olsen, B. T. "Economics of residential broadband access network technologies and strategies," *IEEE Network* **11**:1, 51–57 (Jan./Feb. 1997).
8. Markoff, J. "Old man bandwidth," *The New York Times* Dec. 8, 1997, pp. D1, D13.

Appendix A

Abbreviations

API	Application Programming Interface
ASN.1	Abstract Syntax Notation One
ATDM	Asynchronous Time-Division Multiplexing
ATM	Asynchronous Transfer Mode
CAD	Computer-Aided Design
CAROT	Centralized Automatic Reporting on Trunks
CGI	Common Gateway Interface
CMIP	Common Management Information Protocol
CO	Central Office
CORBA	Common Object Request Broker Architecture
CRV	Call Reference Value
DCOM	Distributed Common Object Model
DIP	Dedicated Inside Plant
DOP	Dedicated Outside Plant
xDSL	Digital Subscriber Line
ADSL	Asymmetric DSL
CDSL	Consumer DSL
HDSL	High-bit-rate DSL
RADSL	Rate-adaptive DSL
SDSL	Symmetric DSL
	Splitterless DSL
VDSL	Very-high-bit-rate DSL
ECCR	Exchange Cable Customer Record
EOC	Embedded Operations Channel
FTTC	Fiber to the Curb
GDMO	Guidelines for the Definition of Managed Objects
GIS	Geographic Information System
GUI	Graphical User Interface
HDT	Host Digital Terminal
HFC	Hybrid Fiber Coax

HTML	Hypertext Markup Language
IN	Intelligent Network
IP	Internet Protocol or Inside Plant
ISDN	Integrated Services Digital Network
LAN	Local Area Network
LDS	Local Digital Switch
LEC	Local Exchange Carrier
LMOS	Loop Management Operations System
MDF	Main Distributing Frame
MIB	Management Information Base
NIU	Network Interface Unit
OA&M	Operation, Administration, and Maintenance
OMNIpoint	Open Management Interoperability Points
ONU	Optical Network Unit
OP	Outside Plant
OS	Operating System (executive that manages computer)
OSI	Open Systems Interconnection
OSS	Operations Support System (application that manages network)
OSSP	Operations Systems Strategic Plan
PCS	Personal Communication Services or Systems
POTS	Plain Old Telephone Service
PVC	Permanent Virtual Channel
QoS	Quality of Service
SDH	Synchronous Digital Hierarchy
SDV	Switched Digital Video
SNA	System Network Architecture
SNMP	Simple Network Management Protocol
SOHO	Small Office/Home Office
SONET	Synchronous Optical Network
SOP	Service Order Processor
SQL	Standard Query Language
SVC	Switched Virtual Circuit
TCP	Transmission Control Protocol
TL1	Transaction Language 1
TMN	Telecommunications Management Network
TN	Telephone Number
TSI	Time/Slot Interchange
UML	Unified Modeling Language
USOC	Universal Service Order Code
WAN	Wide Area Network
WDM	Wavelength-Division Multiplexing

Appendix B

Bibliography of Useful Patents, Books, and Articles

1. PATENTS

Bernstein, L., Assignee: Bell Telephone Laboratories, AT&T Bell Laboratories. "Arrangement for dynamically identifying the assignment of a subscriber telephone loop connection at a serving terminal," U.S. Patent 5,355,405 (Oct. 11, 1994).

Bernstein, L., Assignee: Bell Telephone Laboratories, AT&T Bell Laboratories. "Routing to intelligence," U.S. Patent 5,390,169 (Feb. 14, 1995).

Bernstein, L., Assignee: Bell Telephone Laboratories, AT&T Bell Laboratories. "Routing to intelligence," U.S. Patent 5,392,277 (Feb. 21, 1995).

Billinger, R. J., Dotter, L. K., Gasaway, T. D., Herrick, D. W., and Johnson, S. W., Assignee: American Telephone and Telegraph Company and AT&T Information Systems. "Inter-exchange carrier access," U.S. Patent 4,769,834 (Sept. 6, 1988).

Boyle, G. C., Assignee: AT&T Bell Laboratories. "Time-ordered data base," U.S. Patent 4,646,229 (Feb. 24, 1987).

Butler, T., Hacker, L. K., Jadhav, S. B., Jessup, A. B., Keffer, R. J., Lamm, I. W., O'Brien, M. A., and Weber, C. C., Assignee: Sprint Communication Company, "Providing communications services in a telecommunications network," U.S. Patent 5,528,677 (Jan. 18, 1996).

Daugherty, T. H., DeBruier, D., Greenberg, D. S., Hodgon, D. J., and Murphy, D. J., Assignee: AT&T Corp. "Communications access network routing," U.S. Patent 5,381,405 (Jan. 10, 1995).

Daugherty, T. H., DeBruier, D., Greenberg, D. S., Hodgon, D. J., and Murphy, D. J., Assignee: AT&T Corp. "Establishing connections in a communications access network," U.S. Patent 5,386,417 (Jan. 31, 1995).

Ferror, R. S., Goldstein, A. J., and Nazif, Z. A., Assignee: AT&T Bell Laboratories. "Hyperedge entity-relationship data base systems," U.S. Patent 4,479,196 (Oct. 23, 1984).

Harper, M. E., Robson, S. L., and Combs, C. W., Assignee: Bell Atlantic Services. "Provisioning a public switched telephone network," U.S. Patent 5,491,742 (Feb. 13, 1996).

Harper, M. E., Robson, S. L., and Combs, C. W., Assignee: Bell Atlantic Services. "Provisioning a public switched telephone network," U.S. Patent 5,416,833 (May 16, 1995).

Maurer, K., Castan, M., Cousin, T., Spagnola, G., Daley, K. M., Keegan, M., and Smith, T., Assignee: Bell Atlantic Network Services. "Method and system for remotely activating/changing subscriber services in a public switched telephone network," U.S. Patent 5,619,562 (April 8, 1997).

Sanders, R. W., Assignee: Circuit Path Network Systems. "Cell slot assignment," U.S. Patent 5,502,723 (March 26, 1996).

2. GENERAL INTEREST

Caruso, J. "HP adds config/change management to OpenView offerings," *Network Management Systems & Strategies* **6**(22) (Nov. 15, 1994).

Frisch, I. T., Malek, M., and Panwar, S. *Network management and control*, Vol. 2 (Plenum Press, New York, 1994).

Goldberg, A., and Rubin, K. S. *Succeeding with objects: Decision frameworks for project management* (Addison–Wesley, Reading, MA, 1995).

Goodman, D. J. *Wireless personal communications systems* (Addison–Wesley, Reading, MA, 1998).

IEEE Network **12**: 4 (July/Aug. 1998). Special Issue PCS Network Management.

Korson, T., and McGregor, J. D. "Understanding object-oriented: A unifying paradigm," *Communications of the ACM* Sept. 1990, 41–60.

McGregor, J. D., and Sykes, D. A. *Object-oriented software development: Engineering software for reuse* (Van Nostrand–Reinhold, Princeton, NJ, 1992).

Pugh, W., and Boyer, G. "Broadband access: Comparing alternatives," *IEEE Communications Magazine* Aug. 1995, pp. 34–46.

Schneiderman, R. *Future talk: The changing wireless game* (IEEE Press, Piscataway, NJ, 1997).

Taylor, D. *Object-oriented technology: A manager's guide* (Addison–Wesley, Reading, MA, 1990).

3. ANALYSIS AND DESIGN METHODS

Booch, G. *Object oriented analysis and design with applications* (Benjamin-Cummings, Redwood City, CA, 1994).

Wilkinson, N. M. *Using CRC cards* (SIGS Books, New York, 1995).

Wirfs-Brock, R. J., Wilkerson, B., and Wiener, L. *Designing object-oriented software* (Prentice–Hall, Englewood Cliffs, NJ, 1990).

4. DESIGN

Ash, G. R., and Chang, F. "Management control and design of integrated networks with real-time dynamic routing," *Journal of Network and Systems Management* **1**(3) (Sept. 1993).

Gacek, C., Abd-Allah, A., Clark, B., and Boehm, B. "On the definition of software system architecture," *ICSE 17 Software Architecture Workshop* (Center for Software Engineering, University of Southern California, Los Angeles, April 1995).

Gamma, E., Helm, R., Johnson, R., and Vlissides, J. *Design patterns: Elements of reusable object-oriented software* (Addison–Wesley, Reading, MA, 1995).

IEEE Software **12**(6) (Nov. 1995).

Shaw, M., and Garlan, D. *Software architecture: Perspectives on an emerging discipline* (Simon & Schuster, New York, 1996).

5. CODING IN C++

Barton, J. J., and Nackman, L. *Scientific and engineering C++* (Addison–Wesley, Reading, MA, 1995).

Coplien, J. O. *Advanced C++ programming styles and idioms* (Addison–Wesley, Reading, MA, 1992).

Mancl, D., and Havanas, W. "A study of the impact of C++ on software maintenance," *Proceedings IEEE Conference on Software Maintenance* (Nov. 1990), pp. 63–69.

Meyers, S. *Effective C++: 50 specific ways to improve your programs and designs* (Addison–Wesley, Reading, MA, 1992).

Murray, R. B. *C++ strategies and tactics* (Addison–Wesley, Reading, MA, 1993).

6. HUMAN FACTORS

Beatty, J. *The world according to Peter Drucker* (The Free Press, New York, 1998).

Computer Science and Telecommunications Board, and National Research Council. *More than screen deep: Toward every-citizen interfaces to the nation's information infrastructure* (National Academy Press, Washington, DC, 1997).

Csikszentmihalyi, M. *Flow: The psychology of optimal experience* (Harper & Row, New York, 1990).

Goleman, D. *Working with emotional intelligence* (Bantam Books, New York, 1998).

Pinker, S. *How the mind works* (Norton, New York, 1997).

Zuboff, S. *In the age of the smart machine* (Basic Books, New York, 1988).

Index

Access network, 1, 12
   intelligent controller, 24
   modernization formula, 169–172
Adapter/collector systems, 135
ADSL, 17, 19
   flaws, 57
   DSL options, 86–89
   technology, 109
Akyildiz, I. F., 45
Alarm processing, 76–77, 112
Architecture
   client/server, 35, 104
   distributed, 86
   Ethernet, 74
   functional, 112–115
   LAN, 34, 69, 74, 76, 98–99, 101
   LEC, 77
   network management, 34
   system integrity, 117–118
   technical issues needing resolution, 38–39
   WAN, 34, 101
ATDM, 73
ATM
   broadband readiness, 82
   design, 42–43
   detailed workings of, 74
   engineering growth, 37
   Fraser, 73
   funnel factor, 22
   overhead, 98
   plug and play, 2
   TCP/IP, 74, 99
   TCP/IP utilization, 25
   trial, 107
   unbundling, 36
ATT, 67, 150
   National Network Control Center, 77
   objects in memory, 120
Autodiscovery, 34, 56–57, 62

Bachman, D., 69
Bell Laboratories, 117
Bell System Standards, 119
Bellcore, 17
   OSSP, 96
   TL1, 102
   standards, 119
Bloomberg, 148
Boehm, Barry, 123
Boeing, 148
Bouloutas, A. T., 65
Boyer, G., 16
Broadband
   assignment/installation, 59
   best bets, 75
   call path, 59
   provisioning, 60
   readiness, 78–83
   traffic flow, 69
Brooks, Fred, 75

Cable companies, 6, 46, 87, 107–108, 166
Capital improvements, 157
CAROT, 3
Casulli, D. L., 4
Cerf, Vinton
   Internet model, 23
   TCP, 6
CGI, 137
Churn, 164–165
CMIP, 99–101, 106, 134
Configuration management, 125–126
   database, 139
Congestion
   Internet, 74
   preventive control, 25–27
   reactive control, 25–27
CORBA, 122, 134
Customer care, 151

DEMCO